D0385684

CATALYSTS FOR CHANGE

WILEY SERIES IN SYSTEMS ENGINEERING

Andrew P. Sage

ANDREW P. SAGE and JAMES D. PALMER
Software Systems Engineering

WILLIAM B. ROUSE
Design for Success: A Human-Centered Approach to Designing Successful Products and Systems

LEONARD ADELMAN
Evaluating Decision Support and Expert System Technology

ANDREW P. SAGE
Decision Support Systems Engineering

YEFIM FASSER and DONALD BRETTNER
Process Improvement in the Electronics Industry

WILLIAM B. ROUSE
Strategies for Innovation: Creating Successful Products, Systems and Organizations

ANDREW P. SAGE
Systems Engineering

LIPING FANG, KEITH W. HIPEL, and D. MARC KILGOUR
Interactive Decision Making: The Graph Model for Conflict Resolution

WILLIAM B. ROUSE
Catalysts for Change: Concepts and Principles for Enabling Innovation

CATALYSTS FOR CHANGE

Concepts and Principles for Enabling Innovation

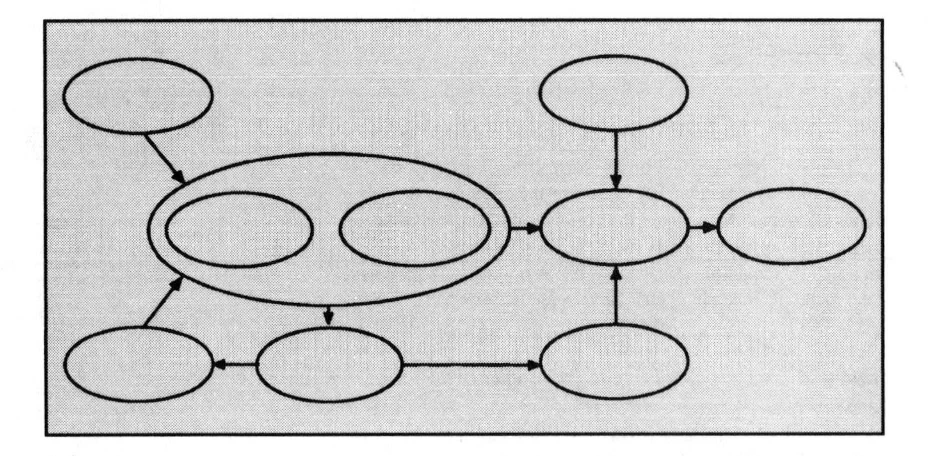

WILLIAM B. ROUSE

A Wiley-Interscience Publication
John Wiley & Sons, Inc.
New York / Chichester / Brisbane / Toronto / Singapore

Library of Congress Cataloging in Publication Data:

Rouse, William B.

Catalysts for change : concepts and principles for enabling innovation / William B. Rouse.

p. cm. — (Wiley series in systems engineering)

"A Wiley-Interscience publication."

Includes bibliographical references and indexes.

ISBN 0-471-59196-3

1. Technological innovations—Management. 2. Industrial management—Social aspects. 3. Systems engineering—Psychological aspects. 4. Export marketing—Management—Social aspects. 5. Intercultural communication. 6. Competition, International. I. Title. II. Series.

HD45.R754 1993 93-364
CIP

Printed in the United States of America

10 9 8 7 6 5 4 3 2 1

Preface

I have spent the past 25 years analyzing, modeling, designing, and evaluating complex systems associated with operating and maintaining a variety of vehicle and process systems. For roughly the last half of this period, I have led a small company that provides software products and engineering services in these areas.

In the mid-1980s, we began to focus on the process whereby products and systems are designed. Our concern, and that of our customers, was that the typical process was rather ad hoc and the results too unpredictable. An extensive development and evaluation effort led to the product realization process described in my book *Design for Success: A Human-Centered Approach to Designing Successful Products and Systems* (Wiley, 1991). The comprehensive methodology in this book focuses on the life cycle of product planning, design, and use.

As we employed this methodology in our company, and helped many customers to adopt all or parts of this methodology, we came to realize that market innovations require more than just good products, systems, and services. Also required is an enterprise that can plan, design, manufacture, sell, and support its offerings to the market.

Innovation involves getting the market to change—to accept and use your products and systems rather than those of competitors. While a product or system may embody one or more creative inventions, without an appropriate enterprise these inventions are unlikely to become market innovations. This realization led to the development and evaluation of a comprehensive, yet relatively simple, methodology for business planning and management that is discussed in my book *Strategies for Innovation: Creating Successful Products, Systems and Organizations* (Wiley, 1992).

For the past five years, part of my job has involved training and assisting our staff and our customers in the use of the concepts, principles, methods, and tools in *Design for Success* and *Strategies for Innovation*. These efforts have been with enterprises ranging from Fortune 500 technology-based companies in the United States, to government agencies in the United States and abroad, to small start-up companies in developing countries. Thus, these experiences have crossed countries, ethnic and corporate cultures, types of institutions, stages of economic development, and often more time zones than I like to remember.

This diversity of experiences has led to another fundamental realization. Beyond the comprehensive methodologies in *Design for Success* and *Strategies for Innovation* there is a somewhat subtle and missing set of ingredients. Having effective and efficient methods for product design and business planning is an important competitive advantage. However, central catalysts in successful use of these methods are the models whereby one understands the world in which the well-planned enterprise tries to sell its well-designed products.

In this book, *Catalysts for Change: Concepts and Principles for Enabling Innovation*, I elaborate a deceptively simple model of the ways in which perceptions are formed and decisions are made. This Needs–Beliefs–Perceptions Model emerged from the aforementioned experiences in situations where needs and beliefs differed substantially from *apparent* mainstream U.S. needs and beliefs. I underscore the word apparent because now that the model has been formalized, I have come to see more diversity of needs and beliefs in the United States than I had before.

This diversity, both in the United States and globally, can be viewed as a problem. However, as I show in this book, recognition and understanding of this diversity also provide opportunities. Some of these are business opportunities, where knowledge of the catalysts for change can provide a competitive advantage. Other opportunities are more societal in nature, whereby recognition of the true underpinnings of sociotechnical disputes and political conflicts can lead to creative new possibilities for resolution.

This book has benefited tremendously from rich interactions with many hundreds of people from a variety of countries, cultures, and sides of issues. While my role in many of these interactions has often been that of teacher and coach, I have usually been just as much a student. Customers for our software products and engineering services, as well as customers for my seminars and consulting, provided sundry insights, pointers, and criticisms. The breadth and robustness of our concepts, principles, methods, and tools are due, in no small part, to these countless contributions.

In particular, I am indebted to Andy Sage for his continued enthusiasm and support throughout the process of developing this book, as well as the earlier two books. Russ Hunt, my partner in the Search Technology "experiment," has also been a continued supporter and contributor to the evolution of these books. I greatly appreciated comments and suggestions by my colleagues Bill Cody and John Hammer as this book emerged.

Many of the insights in this book were motivated by extensive traveling in the past few years. Joe Bitran has facilitated my learning about the Middle East and Africa. Johanan Bithan, Emanuel Eisenberg, Emanuel Sharon, and Haim Yawetz have been my teachers in Israel. Coen Bester, Dumisani and Victoria Sikhakhana, Ferdie Stark, Louis Van Biljon, and Jan Von Tonder have been my instructors in South Africa. Ben Zulu introduced me to Zimbabwe. Howard Atherton and Iain Bitran have helped me to learn about the United Kingdom. Klaus Jamin has provided insights into Germany. Kinji Mori has been my teacher in Japan. Shang-Hwa Hsu and Yuan-Liang Su have taught me about Taiwan. Eugenio Poma was my teacher in Bolivia.

I am also indebted to Bonnie Sikorski, who helped me to create this volume. Sueann Gustin has continued to coordinate seminar/workshop offerings, as well as preparation of related proposals. These efforts have been central to creating, testing, and refining the material in this book.

WILLIAM B. ROUSE

Atlanta, Georgia
July 1993

Contents

CATALYSTS FOR CHANGE

Chapter 1

Introduction

The process of innovation—creating change—involves dealing with an increasingly complex world. Economic competition, political turmoil, and environmental concerns are now global issues. The sense and reality of globalization are rapidly growing, abetted significantly by increasing communication via, for example, Cable Network News' 30-minute updates of worldwide trends and events.

While increasing complexity can make innovation more difficult, it also can provide more opportunities for those who can best cope with this complexity. This book provides a variety of concepts and principles, many heuristics, and a few speculations that can enable innovation or, in other words, serve as catalysts for change. The overall approach to coping with complexity emphasizes understanding the basis for innovation—why people accept or reject change.

A primary concern of this book is provision of practical insights that will enable solving real problems. This type of information is more readily understood in the context of examples or case studies. Consequently, this book focuses on the four archetypical innovation problems listed in Figure 1.1.

ARCHETYPICAL INNOVATION PROBLEMS

The first archetypical problem concerns selling products, systems, and services in the global marketplace. Innovation in the global marketplace requires understanding cultural differences (e.g., Western vs. Eastern) and class differences (e.g., developed vs. developing countries). Depending on the nature of the innovation,

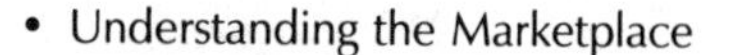

- Understanding the Marketplace
- Enabling the Enterprise
- Settling Sociotechnical Disputes
- Resolving Political Conflicts

FIGURE 1.1. Archetypical innovation problems.

it also may require understanding subtle domain differences (e.g., military aviation vs. commercial aviation) or substantial domain differences (e.g., aerospace electronics vs. consumer electronics).

The essence of this archetypical problem is determining what products, systems, or services various segments of the marketplace will value and how much they will be willing to pay for this value. This is a standard business problem. However, it is very much complicated by globalization. The nature of these complications is discussed in general in the next few chapters, and in detail in Chapter 6.

The second archetypical innovation problem concerns designing an enterprise for global competition. Innovations or changes in the nature of an enterprise involve dealing with power and values in corporate hierarchies, resolving turf battles associated with change, and creating/maintaining appropriate labor relations as changes unfold.

For an enterprise to market and sell products, systems, and services in other cultures—either business, ethnic, or class cultures—it is usually necessary for the enterprise to change its own culture. From personal experience, I know that such changes can be very difficult to accomplish. Well-intentioned employees balk and renege without realizing it. The "good old days" often get better and better as they recede into the past.

The essence of this archetypical problem is determining how the enterprise's culture should change, what means are appropriate for making these changes, and how to deal with people's likely reactions to these changes. This problem presents subtleties that are quite different from those associated with understanding the marketplace. However, as the next few chapters illustrate, both types of problem can be understood within the same general framework. The problem of enabling the enterprise is discussed in detail in Chapter 7.

The third archetypical innovation problem involves settling sociotechnical disputes. Examples include safety of nuclear power, trade-offs between economic development and environmental protection, and the sources of trade imbalances. The nature of such disputes can range from disagreements about facts (e.g., nuclear power), to disagreements about priorities (e.g., economics vs. environment), to difficulties arising due to past differences of priorities (e.g., trade imbalances).

The essence of this archetypical problem is determining how different stakeholders view the dispute and why they have these views. Disagreements about solutions are typical—otherwise, there would not be a dispute. However, below the surface of apparent facts and priorities, there often are more subtle disagreements which, when understood, can be creatively resolved. The means for getting "below the surface" are described in the next few chapters. The problem of settling sociotechnical disputes is discussed in detail in Chapter 8.

The fourth and final archetypical innovation problem involves resolving political conflicts. Examples include ongoing conflicts in the Middle East and South Africa. Such conflicts are typically laden with territorial, economic, ethnic, and religious disputes. It is often difficult to get the conflicting parties to agree to attempt to resolve the conflict peacefully.

The essence of this innovation problem is determining how the different conflicting parties view the conflict and why these views are held. Getting below the surface of these perceptions may enable transforming an inherent win–lose situation into a possible win–win situation. In the next few chapters, means for such an analysis are elaborated. In Chapter 9, the problem of resolving political conflicts is discussed in detail.

CENTRAL ISSUES

It seems quite reasonable to assert that the four types of problems just described are archetypical innovation problems in the sense that "old" solutions are unlikely to succeed. However, this assertion begs the question, "What is innovation?" In *Strategies for Innovation* (Rouse, 1992), I noted that innovation is the introduction of change via something new. The key words here are "change" and "new."

In the context of the four archetypical problems, the goal is to introduce new products, systems, and services; new organizational functions and roles; and new solutions of disputes and conflicts. Not only do we want to introduce such things—we want people to change past patterns of behavior and both accept and endorse these new things. Clearly, the acceptance and endorsement of change strongly depend on the nature of what is new, as well as how it is introduced.

In *Design for Success* (Rouse, 1991), as well as *Strategies for Innovation*, I introduced and illustrated the roles of the seven issues in Figure 1.2. I have found these issues to be central to creating and introducing product and system innovations. This set of issues can be partitioned into two sets.

The four issues toward the bottom of this list (i.e., evaluation, demonstration, verification, and testing) are primarily engineering oriented. For a solution to be successful, it is *necessary* to address these four issues appropriately. However, resolution of these issues is not *sufficient* for success.

The three issues toward the top of the list in Figure 1.2 (i.e., viability, accept-

Viability	→	Are the Benefits of the Solution Sufficiently Greater than its Costs?
Acceptability	→	Do Organizations/Individuals Adopt the Solution?
Validity	→	Does the Solution Solve the Problem?
Evaluation	→	Does the Solution Meet Stated Requirements?
Demonstration	→	How Do Observers React to the Solution?
Verification	→	Is the Solution Put Together as Planned?
Testing	→	Does the Solution Run, Compute, Etc.?

FIGURE 1.2. Central issues.

ability, and validity) are primarily market oriented. These issues are concerned with the extent to which a solution solves the real problem, solves it in an acceptable way, and solves it in a way that is worth the cost. Appropriate resolution of these three issues is sufficient for a solution to be successful, assuming that the four engineering-oriented issues have been dealt with appropriately.

My experience is that it is best to first address the three market-oriented issues, and then address the four engineering-oriented issues. In this way, resources are not invested in testing, verifying, and so on, manifestations of concepts that are inherently invalid, unacceptable, and/or not viable. Unfortunately, my experience also has been that the natural tendency of technology-based enterprises in particular, and enterprises in general, is to focus on engineering or technical issues first and concern themselves with the market later. The result is many inventions, but very few innovations.

The primary focus of this book is viability, acceptability, and validity. My central thesis is that innovation can be enabled by understanding why people perceive (or do not perceive) solutions to be viable, acceptable, and valid. As a precursor to developing this thesis, it is important to elaborate on the definitions of these three issues.

Stakeholders

A central element in defining and measuring viability, acceptability, and validity concerns the different perspectives from which these issues are viewed. For the four types of problems pursued in this book, there are usually multiple stakeholders—people or classes of people who have a stake in the solution of a problem.

To understand the marketplace—our first archetypical problem—one should consider stakeholders such as customers, users, maintainers, and regulators. Enabling the enterprise should involve consideration of employees, customers, and investors. Settling sociotechnical disputes and resolving political conflicts inherently involve multiple groups of stakeholders with, at least on the surface, diametrically opposed positions.

Much of the discussion in this book focuses on understanding the basis for different points of view. If we understand why different stakeholders perceive viability, acceptability, and validity differently, the true nature of differences can be identified. Such insights can enable innovative solutions.

Viability

Referring to Figure 1.2, viability is concerned with the question, "Are the benefits of the solution sufficiently greater than its costs?" It is quite possible to have a valid and acceptable solution that is not viable, simply "not worth it."

How do stakeholders characterize benefits? Customers may be concerned with expanded functionality, high performance, or an appealing image. Employees typically want interesting work, good pay, and pleasant surroundings.

For sociotechnical disputes, one set of stakeholders may want increased economic development while another may desire increased environmental protection. In all likelihood, both sets of stakeholders will want both benefits, but disagree on priorities.

With regard to political conflicts, all groups of stakeholders may want territory and economic opportunities. Typically, however, territory and opportunities are scarce resources. On the surface, therefore, conflicts may appear unresolvable.

How do stakeholders characterize costs? Customers are usually concerned with purchase price, likely maintenance costs, and efforts required to learn to use and subsequently use a product. Employees typically focus on effort required, stress involved, hours necessary, and potential risks and uncertainties.

Costs of potential solutions of sociotechnical disputes include not only money, but also risks and uncertainties relative to, for example, the local economy and the environment. Resolution of political conflicts can involve costs that include money, dislocation of people, and disturbance of political balances.

Acceptability

This issue concerns the question, "Do organizations/individuals adopt the solution?" Solutions that are perceived to be both viable and valid may also be perceived as unacceptable because they require too much change of habits, do not fit into the current ways of doing things, and do not function as expected.

Stakeholders may question the credibility of technologies likely to underlie solutions. They may also question the feasibility of any necessary procedural or organizational changes. Perceived risks and uncertainties may also undermine acceptability.

For customers, a viable and valid solution may be unacceptable because it is novel. Employees may find solutions unacceptable if they affect their roles and status in ways that conflict with their self-images. Disputants on one side of a sociotechnical dispute may find inherent technological risks unacceptable. Adversaries in a political conflict may not accept particular gains by their counterparts.

Validity

This issue concerns the question, "Does the solution solve the problem?" This, of course, leads to the question, "What is the problem?"

A primary difficulty with this question is that stakeholders are often too willing to immediately conclude what the problem is. Further, they often have already decided upon the solution. Often, great progress can be made by getting stakeholders to focus on defining the problem rather than debating solutions.

Beyond defining the problem, it is important to determine how one would know if the problem was solved. In some situations, technically defined and objective measures are available and acceptable to all stakeholders. More often, however, subjective measures are predominant.

Customers may be concerned with who else has bought the product or who has endorsed it. Employees' perceptions of validity may be strongly related to the opinions of their peers and the extent to which organizational changes do not clash with their values and priorities. Participants in disputes and conflicts may heavily weigh statements by opinion leaders—on their side of the issue—in assessing validity.

Discussion Space

Summarizing briefly, thus far I have discussed four archetypical innovation problems and several central issues associated with solving these problems. Three of these issues—viability, acceptability, and validity—have been emphasized as particularly important.

Figure 1.3 juxtaposes primary stakeholders (for each type of problem) and the three primary issues to yield a "discussion space" that is central to this book. It is important to note that only primary stakeholders are indicated in this figure. The archetypical problems typically involve multiple classes of stakeholders which are considered in the chapter associated with each problem.

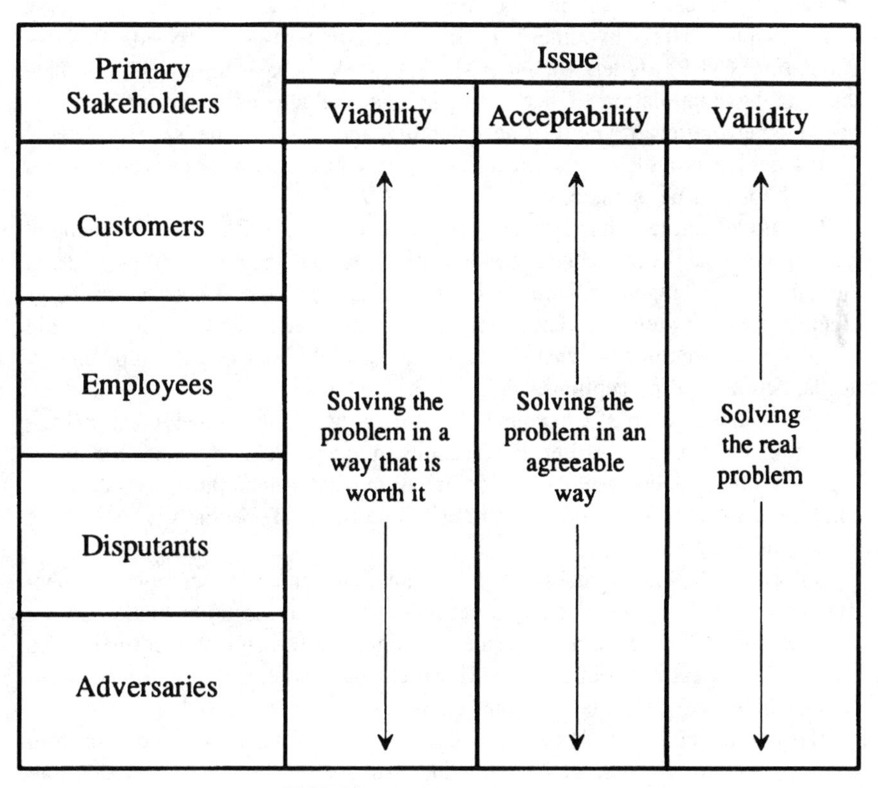

FIGURE 1.3. Discussion space.

Within the discussion space, the concern is with whether or not stakeholders perceive solutions as viable, acceptable, and valid. Of particular interest in this book is why their perceptions are positive or negative. Also of interest are individual, organizational, and societal perspectives on the basis of these perceptions.

FRAMEWORK FOR ANALYSIS

Consideration of the basis of people's perceptions leads to a hypothesis that is central to the line of reasoning developed in this book:

> *People's knowledge and the stated facts are not the sole determinants of perceptions. People's needs and beliefs affect what knowledge is gained, what facts are sought, and how both are interpreted.*

This hypothesis is elaborated and evaluated in subsequent chapters, particularly Chapters 2–5. The implications of this hypothesis are illustrated in the discussion of the archetypical innovation problems in Chapters 6–9.

In Chapter 2, this hypothesis is formalized in terms of a Needs–Beliefs–Perceptions (NBP) Model. The NBP Model provides a basis for a NBP Template that enables compilation of needs and beliefs likely to influence positive and negative perceptions of viability, acceptability, and validity. The NBP Model and NBP Template provide an analytical framework within which the central concerns in this book can be pursued.

The model and template are elaborated in two ways. In Chapters 2–5, general knowledge about the influence of needs and beliefs on perceptions is reviewed and compiled by drawing upon literature from behavioral and social sciences, management, history, and religion. The overall goal is to evaluate and compile—within the model and template—what is known in general about relationships among needs, beliefs, and perceptions.

Discussions in each of Chapters 6–9 focus on one of the archetypical innovation problems discussed earlier. By focusing on one particular type of problem, it is possible to draw upon more specific literature sources than those considered in Chapters 2–5. This material is used to elaborate problem-specific versions of the model and template.

Unfortunately, but inevitably, the literature is inadequate for completely specifying the model and template. It is necessary, therefore, to go beyond concepts and principles gleaned from the results of formal studies and consider heuristics based on practical experience. A wide variety of heuristics can be drawn from popular literature, particularly in management.

Many of the heuristics discussed in Chapters 6–9 of this book are drawn from my personal experiences. In fact, not surprisingly, the hypothesis, model, and template emerged from my wrestling with problems in my company and those of the many companies with whom I work. Basically, management and engineering experience led me to hypothesize phenomena and causes. My scientific training and experiences caused me to seek confirmation (or disconfirmation) in formal reports of studies of needs, beliefs, and perceptions of individuals, organizations, and societies. Where knowledge was found to be lacking, practical heuristics were gleaned from a wide range of experiences.

It is important to characterize these experiences to provide readers with a basis for judging the relevance of conclusions to their concerns. Reviewing past activities, I determined that my colleagues and I have worked with almost 100 enterprises. Roughly half of this work has been associated with applying the concepts, principles, methods and tools of *Design for Success, Strategies for Innovation*, and this book. The other half involved providing software products or engineering services to these enterprises. Approximately one-third of these activities occurred outside of the United States.

Figure 1.4 indicates the distribution of these activities across market sectors. It is useful to note that each of the almost 100 enterprises was counted only once, regardless of the number of different activities with them and independent of the number of different divisions or organizations within the enterprise with whom we have worked.

From Figure 1.4 it can be seen that my experience is biased toward technology-based enterprises in general, and aviation, computer hardware, computer software, electronics, and control systems in particular. Thus, the relevance of many of my

MARKET SECTOR	PERCENT OF TOTAL
Commercial Aviation	16
Computers and Electronics	16
Power and Processes	16
Government Agencies	16
Technical Services	13
Manufacturing, Mining, Etc.	12
Military Aviation	10

FIGURE 1.4. Involvement across market sectors.

observations to, for example, fast food restaurants and fashion boutiques may be questionable and left to readers to judge.

To summarize, the NBP Model and NBP Template are elaborated first by drawing upon formal reports of study results. Holes remaining are filled in using heuristics drawn from practical experience. The sources of conclusions—empirical or experiential—are always indicated.

The completed models and templates are used as a means for suggesting a variety of *potentially* innovative solutions to the archetypical problems. The nature of these solutions illustrates the value of getting "below the surface" of perceptions and understanding positive or negative reactions in terms of needs and beliefs.

OVERVIEW

The remainder of this chapter is devoted to providing an overall perspective of this book. This includes discussion of the central themes in the book and how each chapter contributes to the line of reasoning that is developed around these themes.

There are three central themes in this book. The first theme is

> *Understanding the nature of viability, acceptability, and validity in any particular context is key to selling products, changing organizations, or solving problems in that context.*

This introductory chapter has elaborated this theme. Chapters 2–5 pursue this understanding. Chapters 6–9 illustrate the value of this understanding within the context of the four archetypical innovation problems. Chapter 10 integrates these conclusions.

The second theme is concerned with formalizing this understanding and providing an analytical framework within which this understanding can be operationalized:

> *Understanding the relationships among needs, beliefs, and perceptions will provide deep understanding of viability, acceptability, and validity.*

As noted earlier in this chapter, perceptions related to these three central issues depend on more than just people's knowledge and the stated facts—their needs and beliefs play key roles in this process. Chapters 2–5 discuss the nature of needs and beliefs from a variety of perspectives including culture, religion, science, and technology. The result is an initial outline (Chapter 2) and subsequent elaboration (Chapter 5) of the Needs–Beliefs–Perceptions Model. Chapters 6–9 focus on applying this model to the four archetypical innovation problems. Chapter 10 integrates the insights gained from these applications.

The third theme is concerned with using the aforementioned understanding to enable innovation:

Innovation can be enabled by designing products, systems, services, organizations, and solutions in general that, at the very least, satisfy needs and do not conflict with beliefs, and potentially may facilitate the constructive evolution of needs, beliefs, and perceptions.

Understanding viability, acceptability, and validity as affected by needs, beliefs, and perceptions can provide a basis for influencing stakeholders to buy products, support organizational growth, and agree to change in general. Chapter 5 introduces mechanisms within the NBP Model to support this process of influencing stakeholders. Chapters 6–9 illustrate specific instances of the use of these mechanisms. Chapter 10 discusses the nature of the resulting innovations.

SUMMARY

This chapter began with a discussion of four archetypical innovation problems:

- Understanding the marketplace
- Enabling the enterprise
- Settling sociotechnical disputes
- Resolving political conflicts

The nature of each of these problems was briefly elaborated.

Discussion next focused on the central issues associated with addressing such problems. It was concluded, based on previous efforts (Rouse, 1991, 1992), that understanding stakeholders' perceptions of the viability, acceptability, and validity of potential solutions is a key element of enabling change.

Next, a glimpse was provided of an analytical framework for pursuing the four archetypical innovation problems. The Needs–Beliefs–Perceptions Model and Template were noted, but not described. The approach adopted for elaborating the NBP Model and Template was described.

Discussion concluded with an overview of the central themes of this book and their relationships with the other chapters. With this foundation laid, we are in a position to begin the exploration for understanding that will provide catalysts for change.

REFERENCES

Rouse, W. B. (1991). *Design for success: A human-centered approach to designing successful products and systems.* New York: Wiley.

Rouse, W. B. (1992). *Strategies for innovation: Creating successful products, systems and organizations.* New York: Wiley.

Chapter 2

Outline of the Model

In *Design for Success* (Rouse, 1991) and *Strategies for Innovation* (Rouse, 1992), I argued that design of successful products, systems, and organizations requires understanding people's perceptions of viability, acceptability, and validity. In this chapter, the sources of these perceptions are considered and elaborated.

The simple model in Figure 2.1 provides a starting point. In this model the results of perception are the inputs to the decision-making process. For example, perceived attributes of alternative products serve as inputs to customers' decision making regarding which product to purchase.

There is considerable debate about how decision making happens. Do people analytically weigh the attributes of alternatives? Or, do they react holistically to the pattern of attributes of each alternative? The answers to these questions are highly dependent on the experience of decision makers and whether they are starting wars, investing in factories, choosing spouses, buying cars, or picking candy bars. For recent reviews of theories and models of decision making, see Sage (1991, 1992) and Klein, Calderwood, and Orasanu (1992).

In this chapter, the focus is on how perceptions are formed rather than how decisions are made. The motivation for this emphasis is quite simple. My experience is that focusing on people's perceptions of your product or solution and assuring that these perceptions are positive—often by modifying the product or solution—usually results in their buying the product or agreeing to the solution.

The conventional model shown in Figure 2.1 represents perceptions as being influenced by a person's knowledge and the information presented to them. Succinctly, people's knowledge gained via education and experience combines with the information available to them—the facts at hand—to yield perceptions.

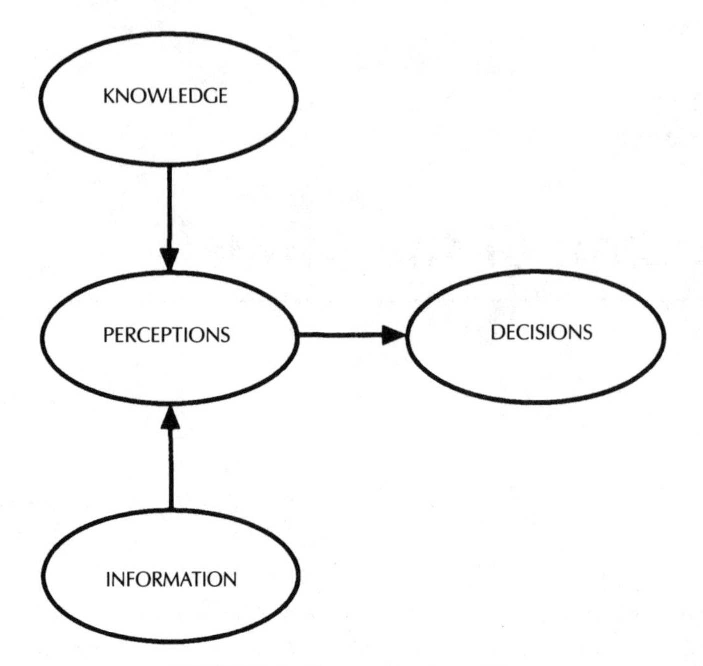

FIGURE 2.1. Conventional model.

Viewed very broadly, this model seems quite reasonable. However, the simplicity of this model can lead to inappropriate conclusions. For instance, in the context of this simple model, if people's perceptions are other than you desire, two courses of action are possible. You can modify the information available and/or educate them to modify and extend their knowledge. In doing so, your goal would be to "correct" misperceptions.

This perspective can lead you to feel that people don't buy your products or accept your solutions because they don't understand. For example, if people really understood nuclear engineering, they would know that nuclear power is safe and they would want nuclear power plants. If people really understood artificial intelligence, they would know that our software is superior and they would buy it.

Obviously, this perspective leads to many unsold products and unacceptable solutions. To enable change—decisions to buy products and accept solutions—a more elaborate model is needed.

NEEDS–BELIEFS–PERCEPTIONS MODEL

Beyond knowledge or "facts in the head," and information or "facts at hand," what influences perceptions? A general answer is that people have a tendency to perceive what they want to perceive. In other words, their a priori perceptions strongly influence their a posteriori perceptions. This tendency is supported by inclinations to gain knowledge and seek information that support expectations and confirm hypotheses (Kahneman, Slovic, and Tversky, 1982).

In this book, these phenomena and a variety of related phenomena are characterized in terms of the effects of people's needs and beliefs on their perceptions. This characterization is depicted in the Needs–Beliefs–Perceptions (NBP) Model shown in Figure 2.2. The remainder of this book deals with elaboration of this model and applying it to the four archetypical innovation problems.

In the context of the NBP Model—in contrast to the conventional model in Figure 2.1—perceptions that disagree with yours are not necessarily incorrect. They are just different. For instance, socialists and capitalists may perceive solutions differently (i.e., in terms of viability, acceptability, and validity) because of differing beliefs about the nature of people and the role of government. As another example, scientists and people who are not technically oriented may have differing perceptions about the impact of technology because of different needs and beliefs concerning the predictability of natural phenomena, desirability of technical solutions, and likely implementations of technologies.

The remainder of this chapter considers the state of general knowledge of needs, beliefs, perceptions, and relationships among these constructs. Prior to this discussion, it is useful to note the likely determinants of needs and beliefs. Figure 2.2 denotes these determinants as nature and nurture.

Nature refers to genetic influences on needs and perhaps beliefs. Physical needs, for example, are obviously affected by inherited characteristics and traits. Characteristics such as stature, tendencies for baldness and poor eyesight, and vulnerabilities to particular diseases appear to be hereditary. Inheritance of psy-

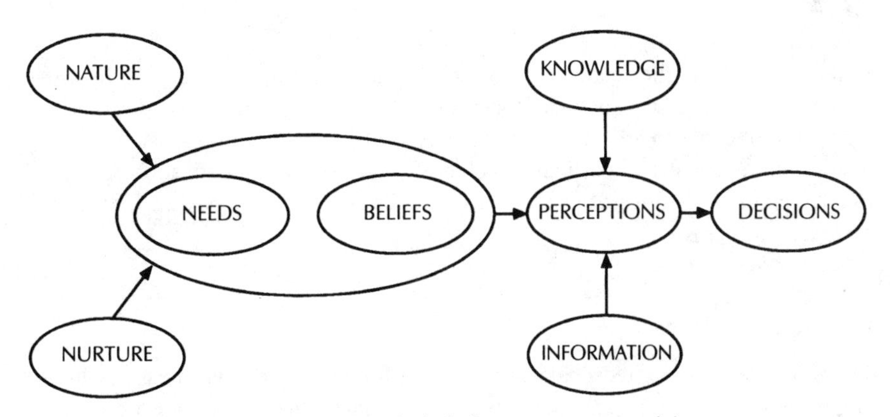

FIGURE 2.2. Needs–beliefs–perceptions model.

chological and social traits is a topic of continuing debate—see Casti (1989), as well as Neubauer and Neubauer (1990), for discussions of the different sides of this debate. Fortunately, the line of reasoning developed in this book does not depend on resolution of the long-standing nature vs. nurture controversy.

Nurture in Figure 2.2 refers to the effects of childhood, education, cultural influences, work experience, economic situation, and so on. Clearly, factors such as work experience and economic situation can have a substantial impact on needs. Similarly, education and cultural influences can greatly affect beliefs.

Nature and nurture are included in the NBP Model to provide means for affecting needs and beliefs. While nature usually is only affected at the snail's pace of evolution, nurture can be more quickly influenced by modifying situations and hence experiences which, thereby, change needs and eventually beliefs, perceptions, and decisions.

Thus, nature provides a very slow means for affecting change, so slow in fact that this topic is not further considered in this book. Nurture provides a faster means. The most expeditious means is that provided by the oval labeled "information" in Figure 2.2. One can modify information to satisfy needs, avoid conflicts with beliefs, and cause perceptions that lead to desired decisions.

While modifying information about your product or solution can be done quickly, its potential impact is limited because needs and beliefs are assumed fixed. Since modifying nurture can change needs and beliefs, the potential impact is much greater. However, this impact emerges more slowly than that due to changes of information. Thus, there is an inherent trade-off between relatively fast and focused changes vs. relatively slow and widespread changes. This trade-off is further discussed in Chapter 5.

NEEDS

In this section and the sections that follow in this chapter, a number of theories from the behavioral and social sciences are reviewed. These theories appear to be broadly applicable to all four of the archetypical innovation problems. Theories that are more specific to particular problems—for instance, theories of organizational change or theories of negotiation—are discussed in the chapters that focus on these problems.

Hierarchy of Needs

The most frequently cited characterization of humans' needs is Maslow's hierarchy shown in Figure 2.3. Maslow (1954) argued that these five needs are activated in a hierarchical fashion. Thus, physiological needs (e.g., food) are

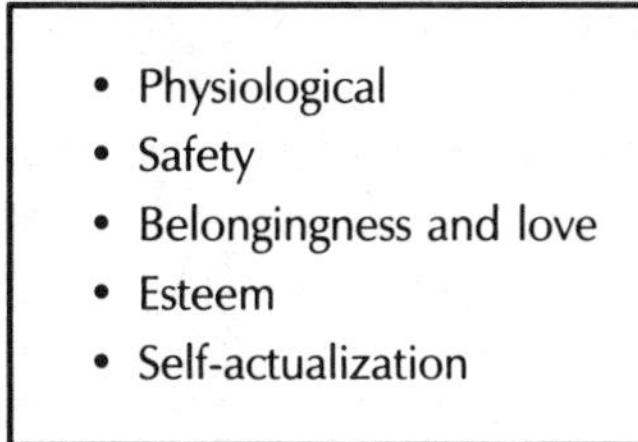

FIGURE 2.3. Maslow's hierarchy of needs (Maslow, 1954).

activated and satisfied before higher-level needs such as safety are activated. Consequently, according to the theory, people will tend, for example, to satiate hunger before they concern themselves with friendship.

Extensive empirical investigations of Maslow's theory have resulted in little support for the hierarchical premises of his theory. In response to this lack of supporting evidence, Alderfer (1972) modified the hierarchy to include just three categories: existence, relatedness, and growth. However, subsequent empirical investigations provided little support for the modified theory.

Lee, Locke, and Latham (1989) reviewed the work of Maslow, Alderfer, and others. They concluded that the primary value of these needs hierarchies is the explanations they provide of what causes people to act. However, these formulations are not very good for predicting specific actions that particular people will take in given situations.

Fortunately, from the perspective of trying to sell products, organizational changes, and solutions in general, it is not at all necessary to assume hierarchies of needs. It is important, however, to recognize the existence of multiple types of need. Understanding the predominant needs in a market sector, or within one's own organization, can help to focus your energies in the most fruitful directions.

I recall a discussion a few years ago with a community leader in the Bolivian Altiplano—the high plains where many impoverished Indians squeeze a subsistence living out of harsh surroundings. While I was there as part of a medical team, my primary interest was in assessing their needs for technical training to operate and maintain vehicles and equipment. The leader told me that his people sorely needed such training. However, he was not sure how much attention the issue would get since many people were primarily concerned with whether or not they would eat that night!

I had a similar discussion in a more recent visit in Bophuthatswana, one of the "homelands" created by South Africa to which millions of blacks were forced to migrate. There I found similar training needs and a variety of training programs. One thing that I had not expected was that many people enrolled in training courses primarily to receive the stipend provided to trainees, not because they expected the training to lead to jobs. In fact, they had very low—and usually correct—expectations of getting jobs.

In the context of organizational change, very recent experiences in shifting my company's emphasis to new markets has resulted in marked changes in employees' conversations concerning needs. My repeated assurances that necessary changes in the company's activities would result in work that was at least as interesting and rewarding as before did not compensate for many employees' feelings that their knowledge and skills might no longer be needed. Uncertainty about the future caused many employees to focus on job security, not job satisfaction. Consequently, we had to focus on the security issue squarely and provide evidence that the future was far less uncertain than many people perceived.

These three examples serve to illustrate the subtlety and multiplicity of needs. They also show how situational attributes can affect the saliency of needs. Therefore, these types of needs are referred to as *situational needs*.

Motives

In contrast to situational needs are *dispositional needs*, or motives that reflect individual dispositions. Murray (1938) pioneered the study of motives and developed a list of 27 "psychogenic needs." Liebert and Spiegler (1970) reviewed Murray's work and discussed the elements of his list in terms of five categories: ambition, defense of status, response to human power, affection between people, and exchange of information.

Most of Murray's needs have received little if any empirical study. Notable exceptions include McClelland's (1987) extensive studies of the four motives listed in Figure 2.4. A primary means for assessing the strengths of people's needs or motives in these studies was analysis of brief stories written by people in response to an experimental manipulation such as being given a performance test.

The results of this long line of research indicated that people with a high need to achieve tend to be those who are strongly oriented toward improving task performance and accepting personal responsibility for performance. They also tend to have a great desire for performance feedback. People with these charac-

- Achievement
- Power
- Affiliation
- Avoidance

FIGURE 2.4. Motives studied by McClelland (1987).

teristics should make good entrepreneurs and people who were already successful entrepreneurs were found to score high in terms of need to achieve.

By analyzing literature from many countries and periods in history, McClelland found that high levels of the need to achieve reflected in this literature were found in countries and time periods associated with increases in entrepreneurial activity and more rapid rates of economic growth. He also found that the literature of reform groups—for example, the early Protestant Reformation prior to institutionalization—reflected high needs to achieve that were associated with economic progress.

The need for power was found to represent a concern for impacting people and things. The power motive leads to openly competitive and assertive behavior, particularly among men. People with a high need for power, coupled with a high degree of internal inhibition (i.e., control of impulses), tend to assume positions of leadership in voluntary organizations and believe in centralized authority, hard work, self-sacrifice, and charity. They make good managers in business enterprises, especially if their need for affiliation is low.

Countries whose popular literature scores high in need for power tend to be more imperial. They collect more of their gross national income in taxes. They spend more of it on defense. While the achievement motive tends to be associated with economic progress, the power motive tends to underlie effective governing.

People with a high need for affiliation tend to perform better when opportunities to relate to other people are present. Such people invest in maintaining personal networks. They emphasize cooperation, conformity, and avoidance of conflict. Because of desires to be liked by everyone, they tend to perform poorly as managers. The need for affiliation is also associated with fear of rejection.

McClelland has also studied a variety of avoidance motives, including fears of failure, rejection, success, and power. Fear of failure—which I have encountered in some employees who were hired because of their perfect or near-perfect college records—has been found to be associated with the parenting practices one was raised with. Fear of rejection has been found to be associated with fear of failure and high need for affiliation. Fear of success tends to be higher for women who fear social rejection based on performance success. Fear of power is associated with concerns that one's assertiveness will lead to rejection by others. While these observations concerning avoidance motives are intriguing, it is important to note that the empirical basis for these conclusions is considerably smaller than that for the other of McClelland's motives.

Summary

In summarizing this discussion of needs, it is important to note the difference between situational and dispositional needs. In order to innovate in new markets or change the nature of organizations, it is important to understand existing and impending situational characteristics as well as prevailing dispositions among key stakeholders. One reason for valuing this distinction is that situations may be changed much more easily than dispositions. Discussions of the archetypical innovation problems in Chapter 6–9 involve consideration of the practical implications of these conclusions.

BELIEFS

Beliefs are those things that are held to be true, that is, consistent with fact or reality. The phenomena of belief involves a wide range of life, including belief that the Earth orbits the Sun, belief that a low-fat diet is good for you, belief that freedom is a right of all humans, and belief in a greater power than humankind. Notice that beliefs are *held* to be true, not necessarily *proven* to be true. Indeed, the whole notion of proof is not uniformly meaningful across the range of illustrative beliefs just listed.

In this section, the nature of beliefs is characterized by reviewing a variety of theories from social science and management, with emphasis on causal models rather than purely descriptive models—see Hammond, McClelland, and Mumpower (1980) for discussions of the latter. Cultural and religious beliefs, per se, are not considered at this point. Chapter 3 is devoted to these topics.

Expectancy Theory

Expectancy theory provides a model of the role of beliefs in decision making (Fishbein and Ajzen, 1975). This theory asserts that the perceived relative attractiveness of various options is related to people's beliefs about the consequences to which each option will lead and their beliefs about the desirability of these consequences.

In situations where probabilities are associated with consequences and/or their level of desirability—in other words, consequences and/or level of desirability are not certain—the theory includes *subjective* probability distributions to discount these risks. *Objective* probability distributions are seldom used, even if they are available, because there is considerable evidence that most people are poor estimators of probabilities (Kahneman, Slovic, and Tversky, 1982). Consequently, to the extent that people use probabilities at all, they inherently use subjective and biased estimates.

The practical implications of expectancy theory are as follows: People react to new products or possible organizational changes, for example, based on what they *believe* will be the consequences of buying the product or implementing the change. Their reactions are also related to what they *believe* will be the level of desirability of the consequences. Finally, any perceived risks associated with consequences and/or level of desirability will cause people to discount the value of the product or the organizational change.

To illustrate, suppose a company were to try to move out of the declining defense-market sector into a growing commercial market sector. This would require significant organizational changes. Would employees be likely to support and facilitate such changes? The answer depends on whether employees believe the company will be successful in the new sector and whether or not they believe the consequences of the success will be desirable for them personally. If they do not believe the company can make this move and/or they do not believe they will like the consequences, implementing change is likely to be difficult.

Thus, your ability to "sell" products, organizational changes, or resolutions of disputes is strongly dependent on stakeholders' beliefs about consequences and levels of desirability. This raises the question of the perceived abilities of your product, change, or resolution to cause desirable consequences. The answer to this question can be pursued by considering the causal relationships that underlie people's beliefs.

Attribution Theory

Attribution theory provides insights into these types of relationship (Kelley and Michela, 1980; Harvey and Weary, 1984). This theory, or collection of theories, is concerned with people's inferences with regard to causation and the consequences of these inferences. Most studies of attribution focus on people's inferences about the causes of their own behaviors or the causes of other people's behavior. When observing other people, such inferences are often prompted by observed behaviors that depart from expected behaviors.

When assessing people's attributions—which typically are assessed by simply asking them—several factors are usually considered. One factor is whether someone is attributing cause to their own actions or those of others. Another factor is the information available upon which to base attributions. A third factor is the needs, desires, and a priori beliefs of those reporting attributions.

Attributions, once assessed, are usually categorized in terms of the nature of the causes attributed to the observed behavior. One category concerns situational vs. dispositional causes. If observed behaviors are believed to be dictated by characteristics of the situation (e.g., an employer may have no choice but to let an employee go), then the attributed cause is termed situational. In contrast, if observed behaviors are believed to be due to personal characteristics of the observed person (e.g., an employer may let an employee go because he or she does not like the worker), then the attributed cause is called dispositional.

A related category of attributions is external vs. internal. This category concerns whether behaviors are caused by external forces or internal motivations. For instance, a salesman's excellent sales performance might be attributed to his luck in being assigned to a region where sales are easy (an external cause) or to his superior sales ability (an internal cause). Similarly, good performance might be attributed to chance (an external cause) or to skill (an internal cause).

One of the most important and interesting findings of research in attribution theory has been what is termed the fundamental attribution error—the tendency to overestimate dispositional causes of behavior. In other words, observed behaviors are often due more to the nature of the situation than to characteristics of the actor. Nevertheless, observers tend to attribute causation to the actor's disposition. Thus, for example, potential organizational changes or resolutions of disputes often run afoul of incorrect attributions of intentions when, in fact, the nature of situations may dictate particular changes or resolutions.

When people are asked to attribute cause to the results of their own behaviors, their attribution errors are somewhat different. They tend to attribute positive results to dispositional causes and negative results to situational causes. When things go well, it's due to their strong abilities and great efforts. When things go poorly, it's due to impossible assignments, inept customers, or bad luck.

Studies have also shown that attributions (or a posteriori beliefs) are affected by a priori beliefs. In particular, people tend to seek information that will confirm their beliefs. If they think an employee is a superior performer, or an adversary is unethical, they will tend to focus on information that supports these beliefs. Consequently, to the extent that they update their attributions, a priori beliefs tend to be strengthened.

Attributions also can lead to self-confirming cycles. Beliefs about other people cause someone to behave toward these people in particular ways, which leads these people to behave in ways that reinforce these beliefs. Numerous studies of male–female relationships have documented this type of cycle. Put simply, your behaviors toward other people tend to determine their behaviors toward you.

It is easy to see how attribution theory might apply to the archetypical innovation problems. Attributed beliefs about other cultures' values, management's intentions, and adversaries' motivations can lead to behaviors, on both sides, that are mutually counterproductive and potentially completely debilitating. By understanding beliefs underlying perceptions and behaviors, it may be possible to break self-confirming cycles of inappropriate attributions.

Mental Models

Attribution theory focuses on people's beliefs about other people's behaviors, intentions, values, priorities, and so on. This is just one class of beliefs. There is a wide range of phenomena about which people adopt or form beliefs, including nature, technology, organizations, and society.

The phrase "mental models" is often invoked to characterize people's understanding of the way the world works, or the way it might work. The mental models construct (or equivalent) has been employed by many disciplines including psychology, management, architecture, and engineering, in some cases for several decades. Thus, this construct is well established.

Nevertheless, it can be difficult to determine what exactly mental models are, and what they aren't. There is a tendency to include in the model virtually everything a person knows. This results in mental models being synonymous with knowledge in general. To be of value, however, the construct must specify particular mechanisms and types of knowledge.

We explored alternative definitions of mental models by reviewing a large number of efforts that employed this construct (Rouse and Morris, 1986). We concluded that mental models are the mechanisms whereby humans are able to generate descriptions of why something exists and its form, explanations of its functioning and what it is doing, and predictions of what it will do in the future. This definition is summarized in Figure 2.5.

This review also led us to consider a variety of issues associated with the formation and use of mental models. Primary among these issues was that of instruction. For any particular task, job, or role, what should people's mental models be? What do they need to know? How can they best gain this knowledge? If tasks are crisply specified, these questions can usually be readily answered. However, for something as ambiguous as someone's role in an organization, these questions are difficult to answer.

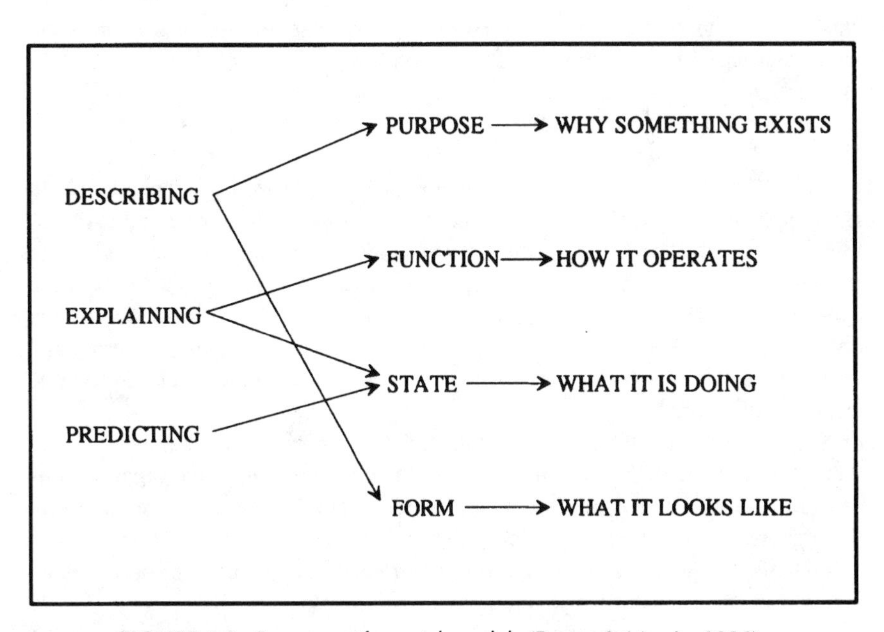

FIGURE 2.5. Purposes of mental models (Rouse & Morris, 1986).

Another issue concerns the extent to which people's mental models can be assessed. If people's mental models are viewed as their attribution of cause relative to some phenomenon, then assessing mental models is equivalent to attributions of attributions of cause. Clearly, there is room for much difficulty in such a process. This topic is considered in detail in Chapter 4 in discussions of fundamental limits in science.

If we now consider the nature of mental models for particular types of phenomena, one class of interest is people's models of space, time, and physical devices (Gentner and Stevens, 1983). People's mental models of the physical world are often intuitive and frequently wrong. For example, their explanations of the physics of bouncing balls and the trajectories of projectiles reveal naive concepts of energy, mass, friction, and so forth, even though, in many cases, the people queried had completed university-level courses in physics. Their everyday experiences appeared to have a much stronger impact on their models than their formal training had. This carries over to mental models of physical devices where novices tend to have fairly simple and very concrete models, while experts are more likely to have more abstract models related to function and purpose.

Johnson-Laird (1983) has studied the mental models that underlie language understanding and communication. He argues that mental models are structural analogs of the world. Mental models describe the structure of events and states of affairs. The implications are, for instance, that communicating the benefits of an organizational change or the merits of a potential resolution of a dispute requires choosing a line of reasoning and particular wording that enables appropriate interpretation within the context of the mental models of recipients of the communication.

We have recently been studying the mental models underlying effective performance of crews on ships (Rouse, Cannon-Bowers, and Salas, 1992). We have posited the existence of team mental models whereby team members are able to form appropriate expectations of the likely behaviors of each other and explanations of behaviors once they are observed. It was hypothesized that inadequate mental models—models with inadequate or incorrect mechanisms for forming expectations and explanations—lead to certain types of team errors. Results of two observational studies found these types of errors to be prevalent. Experimental efforts are now under way to link particular types of model deficiencies to specific instances of error.

Senge (1990) has looked at mental models from an organizational point of view. He concludes that insights into how to change and improve organizations often fail because they conflict with deeply held mental models. To a great extent, this occurs because mental models affect what we see—what information we seek—and, consequently, how we act. The most difficult problem arises when mental models are tacit, below the level of awareness.

The prevalence of this type of problem prompted me to include Chapter 5, "A

Model of the Enterprise," in *Strategies for Innovation* (Rouse, 1992). In working with organizations that are often trying very hard to change, I have found that a point in discussions will be reached where people from marketing and engineering, for example, will argue heatedly about one strategy vs. another. With a little probing, I typically find that marketing's strategy makes sense within its model of the enterprise and engineering's strategy makes sense within its model. However, the two models are incompatible—the two groups are not talking about the same company!

The simple model in Chapter 5 of *Strategies for Innovation* provides a reference point against which different groups can calibrate their models. This almost always leads to a dialogue that uncovers discrepancies among models, many of which are immediately remediated. Thus, making tacit models explicit can be a catalyst that enables change.

The mental models construct provides a means for characterizing people's beliefs about how the world works and how it might work. Such models influence what we look at, what information we seek, what we expect to happen, and how we explain what has happened. Discrepancies among, typically tacit, mental models of stakeholders play a central role in making problems difficult to solve, as is illustrated in later discussions of the four archetypical innovation problems.

Summary

People's beliefs about likely consequences and the level of desirability of these consequences strongly influence their decisions. Beliefs can arise from attributions of causality of observed behaviors. The attribution process is subject to several types of bias—overestimating dispositional causes of behavior, seeking only confirming information, and behaving toward others in ways that causes others to react in manners that support incorrect attributions. The mental models construct provides a means for capturing people's beliefs about how the world works. Discrepancies among mental models, particularly if they are tacit, can lead to lack of communication, poor team performance, and disabled organizations.

PERCEPTIONS

How do needs and beliefs affect perceptions? This is a broad question, covering most of psychology and much of other social sciences. Fortunately, the goals of this book do not require pursuit of a broad answer to this question.

We are concerned with perceptions that influence the decisions associated within the four archetypical innovation problems introduced in Chapter 1 and discussed in detail in Chapters 6–9. These decisions include purchasing a product, "buying in" to organizational change, agreeing to a solution of a sociotechnical dispute, and accepting resolution of a political conflict.

The perceptions of interest are stakeholders' perceptions of viability, acceptability, and validity of products, changes, solutions, and so on. Within the context of the NBP Model, these perceptions are considered using the Needs–Beliefs–Perceptions Template shown in Figure 2.6. This template—the NBP Template—provides a means for accounting for the possible impacts of needs and beliefs or perceptions.

The NBP Template is completed for each stakeholder or class of stakeholders. Explanations of positive or negative perceptions of viability, acceptability, and validity are compiled in the template in terms of needs and beliefs. Comparisons of templates across stakeholders enable identification of common concerns and differences among stakeholders.

This chapter has only provided the rudiments of the knowledge necessary to fill in the template. Chapters 3 and 4 provide a wealth of information about needs and beliefs as they relate to culture, religion, science, and technology. Chapter 5 elaborates the NBP Model. Chapters 6–9 provide much problem-specific information for filling in the template. Chapter 10 discusses the implications of the model and template.

Nevertheless, despite the fact that we are just getting started, consider an example of implementing organizational change due to a company's need/desire to move into new sectors of the marketplace, perhaps due to waning of demands in old sectors. For the sake of illustration, assume that marketing strongly supports

ATTRIBUTE	PERCEPTIONS	BELIEFS	NEEDS
Viability	Positive		
	Negative		
Acceptability	Positive		
	Negative		
Validity	Positive		
	Negative		

FIGURE 2.6. Needs–beliefs–perceptions template.

the change while engineering strongly opposes it. Probing deeper—how this is done is illustrated in later chapters—we discover that marketing perceives the viability, acceptability, and validity of the change quite positively, but engineering is negative relative to acceptability and validity. Engineering agrees that the benefit-to-cost ratios look good, but it does not perceive the change to be acceptable within the company and it does not perceive that it solves the company's problems.

Analyzing engineering's misgivings in terms of expectancy theory, we find that it does not believe that the company can succeed in the new market and it does not believe the consequences are very desirable. In fact, it feels that engineering will lose status in the new market where state-of-the-art technology is not crucial. In contrast, marketing is relieved by the change because it will not have to sell high-priced, hard-to-explain products to a declining market.

From the perspective of attribution theory, we find that engineering has long attributed the company's success to the high level of technical expertise of the engineering staff. In fact, in its interaction with technical staff members in customer companies, the engineering staff has always received high marks for technical sophistication. However, the engineering staff has had almost no contact with end users of the products it designs and develops.

From a mental models point of view, it is realized that the engineering staff has a rather myopic model of the company's relationship with the marketplace. This realization, in combination with the other insights gained in this analysis, leads the company to accompany the organizational change with training that helps engineering staff members to understand and value their somewhat changed roles.

While we did not actually fill in the NBP Template for this example analysis, the "flavor" of such an analysis should be apparent. The template is used in later chapters where analyses are more complex.

SUMMARY

This chapter has introduced the NBP Model and the associated NBP Template. Needs were discussed in terms of a hierarchy of needs and the nature of motives. Beliefs were considered by drawing upon the theories of expectancy and attribution, as well as the construct of mental models. The influences of needs and beliefs on perceptions were discussed in the context of perceptions of viability, acceptability, and validity relative to the four archetypical innovation problems.

At this point, we have just scratched the surface of needs, beliefs, and perceptions. If the goal is to compete globally, we have to understand other cultures—to understand other cultures, we have to understand the world's religions. Similarly, if the goal is to change organizations within technology-based enterprises, we have to understand science and technology in terms of needs, beliefs, and perceptions.

Chapters 3 and 4 consider culture and religion, and science and technology, respectively. Understanding of these two domains leads us back to elaboration of the NBP Model in Chapter 5.

REFERENCES

Alderfer, C. P. (1972). *Existence, relatedness, and growth: Human needs in organizational settings*. New York: Free Press.

Casti, J. L. (1989). *Paradigms lost: Images of man in the mirror of science*. New York: Morrow.

Fishbein, M., and Ajzen, I. (1975). *Belief, attitude, intention, and behavior*. Reading, MA: Addison-Wesley.

Gentner, D., and Stevens, A. (Eds.) (1983). *Mental models*. Hillsdale, NJ: Erlbaum.

Hammond, K. R., McClelland, G. H., and Mumpower, J. (1980). *Human judgment and decision making*. New York: Praeger.

Harvey, J. H., and Weary, G. (1984). Current issues in attribution theory and research. *Annual Review of Psychology, 35*, 427–459.

Johnson-Laird, P. N. (1983). *Mental models: Towards a cognitive science of language, inference, and consciousness*. Cambridge, MA: Harvard University Press.

Kahneman, D., Slovic, P., and Tversky, A. (Eds.) (1982). *Judgment under uncertainty: Heuristics and biases*. Cambridge, UK: Cambridge University Press.

Kelley, H. H., and Michela, J. L. (1980). Attribution theory and research. *Annual Review of Psychology, 31*, 457–501.

Klein, G. A., Calderwood, R., and Oransanu, J. (Eds.) (1992). *Decision making in action: Models and methods*. Norwood, NJ: Ablex.

Lee, T. W., Locke, E. A., and Latham, G. P. (1989). Goal setting theory and job performance. In L. A. Pervin (Ed.), *Goal concepts in personality and social psychology* (Chapter 8). Hillsdale, NJ: Erlbaum.

Liebert, R. M. and Spiegler, M. D. (1970). *Personality: An introduction to theory and research*. Homewood, IL: Dorsey Press.

Maslow, A. H. (1954). *Motivation and personality*. New York: Harper.

McClelland, D. C. (1987). *Human motivation*. Cambridge, UK: Cambridge University Press.

Murray, H. A. (1938). *Explorations in personality*. New York: Oxford University Press.

Neubauer, P. B., and Neubauer, A. (1990). *Nature's thumbprint: The new genetics of personality*. Reading, MA: Addison-Wesley.

Rouse, W. B. (1991). *Design for success: A human-centered approach to designing successful products and systems*. New York: Wiley.

Rouse, W. B. (1992). *Strategies for innovation Creating successful products, systems and organizations*. New York: Wiley.

Rouse, W. B., Cannon-Bowers, J. A., and Salas, E. (1992). The role of mental models in team performance in complex systems. *IEEE Transactions on Systems, Man and Cybernetics, 22,* 1296-1308.

Rouse, W. B., and Morris, N. M. (1986). On looking into the black box: Prospects and limits in the search for mental models. *Psychological Bulletin, 100*, 349–363.

Sage, A. P. (1991). *Decision support systems engineering*. New York: Wiley.

Sage, A. P. (1992). *Systems engineering*. New York: Wiley.

Senge, P. M. (1990). *The fifth discipline: The art and practice of the learning organization*. New York: Doubleday/Currency.

Chapter 3

Culture and Religion

Chapter 2 focused on the psychological basis of needs, beliefs, and perceptions. The discussion, at that point, was relatively context free. However, needs, beliefs and perceptions always occur in a context.

In this chapter, the social and cultural context of needs, beliefs, and perceptions is considered. In order to elaborate this context in a systematic and integrated manner, the discussion focuses on the world's religions. This focus provides an excellent basis for understanding the contextually laden reality of the phenomena captured in the Needs–Beliefs–Perceptions (NBP) Model.

I hasten to note that the goal of this chapter is not to evaluate the veracity or credibility of religions or belief systems. Instead, the panorama of the world's religions provides a social and cultural backdrop that enables insights into differences and similarities among belief systems, as well as how the dynamics of cultures and religions affect belief systems.

Also of interest are organizational cultures and religions as exhibited by a wide variety of enterprises. This topic is discussed in a later section of this chapter. Cultures and religions within science and technology are considered in Chapter 4.

This chapter, as well as the next chapter, provides important background material for Chapters 6–9. Competing globally (Chapter 6) requires an understanding of global business and ethnic cultures. Changing organizations (Chapter 7) requires an understanding of organizational cultures. Resolving sociotechnical and political disputes (Chapters 8–9) requires an understanding of potential cultural components of these disputes.

To provide this background, a brief cultural history of religion is first presented.

This is followed by discussion of the needs, beliefs, and perceptions associated with religions. This chapter concludes by considering organizational cultures and religions.

CULTURAL HISTORY OF RELIGION

Figure 3.1 summarizes the characteristics of the world's major religions. The columns have somewhat different meanings for the different religions, which are clarified in the following discussions.

Hinduism is a religion with ancient roots in India. Its ideal is Brahman, the supreme god. One image or aspect of Brahman is Krishna, a hero that has come to Earth. In addition, Hindus revere millions of other gods, which reflect the infinite aspects of Brahman.

Hindus believe in reincarnation, whereby people continually die and are reborn.

RELIGION	ORIGIN	IDEAL	CENTRAL TEXTS
Hinduism	c. 1500 B.C.	Brahman/Krishna	Upanishads
Buddhism	563 B.C. Siddhartha Gautama	Bodhisattva/ Buddha	Dhammapada
Taoism	604 B.C. Lao Tzu	The Tao	Tao Te Ching
Confucianism	551 B.C. Confucius	Chun-Tzu	Five Classics
Judiasm	c. 1900 B.C. Abraham	God	Torah/Talmud
Christianity	c. 4 B.C. Jesus	God/Christ	Bible
Islam	A.D. 570 Mohammed	Allah	Koran

FIGURE 3.1. The world's major religions.

Virtuous living leads to being reborn in higher and higher castes until one is eventually released from rebirth and united with Brahman. Castes range from the lofty Brahmins, who have special religious duties, to lowly untouchables. While caste once totally dictated one's lot in life, this is no longer true.

Buddhism emerged, in part at least, as a protest against the caste system of Hinduism. Siddhartha Gautama, a wealthy prince in India, rejected his life of luxury and sought to understand the misery and sorrow typified by the caste system. His "enlightenment," achieved after spending 49 days under the Bodhi tree, lead to a life of teaching and the emergence of Buddhism.

His teachings are embodied in the Four Noble Truths, the Noble Eightfold Path, and the Five Precepts, the essence of which includes ethical living, moderation, and withdrawal from the cravings of the everyday world. By living according to these principles, it is believed that one can gain enlightenment, escape reincarnation, and achieve Nirvana.

One who achieves enlightenment becomes a Buddha, the most compelling example of which is Gautama. A Bodhisattva is one who delays entering Nirvana until the whole human race is enlightened. Gautama Buddha, who delayed entering Nirvana for many years after his enlightenment in order to travel and teach, represents the qualities of a Bodhisattva.

Buddhism is prevalent in India, southern Asia, China, Japan, and Korea. Hinayana Buddhism in southern Asia emphasizes individual responsibility and salvation by personal example. Mahayana Buddhism in China, Japan, and Korea espouses salvation by faith and good works.

Buddhism was introduced into Japan in roughly a.d. 700. Since that time, after many ups and downs, it has achieved at least equal status with that of Shintoism, a much more primitive religion typified by animism—a belief that gods are embodied in many, if not all, creatures and objects. Many Japanese honor and frequent both Buddhist temples and Shinto shrines.

I have asked several Japanese friends why people seem to practice two, somewhat inconsistent, religions. The answer has always been, "We are Shinto when we are born and married, Buddhist when we die." My interpretation of this answer is that the here-and-now celebratory nature of Shinto fits well with the joys of birth and marriage, while the abstract, philosophical nature of Buddhism is more helpful for coping with death.

The philosophies of China—Taoism and Confucianism—are focused on ethical living rather than God, the hereafter, and so forth. The emphasis is on the unity of man and nature, with a strong sense of tradition and joy. One is urged to strive for harmony between one's Yin (feminine nature) and Yang (masculine nature).

The Tao—the ideal for Taoists—is the way toward balance, with emphasis on that which is natural, simple, good, and loving. If everyone lived by the Tao, it is believed there would be no ambition for power or war. Friendship and brotherly love would prevail.

Confucius, who followed shortly after Lao Tzu, supported the Tao. He built a practical system of ethics to go with it. This system emphasizes social order and relationships between, for example, husband and wife, and father and son. Of particular importance is respect of children for parents.

Buddhism arrived in China in approximately a.d. 400. Since then, Taoism and Confucianism have had to accommodate the appeal of Buddhism. The result has been an acknowledgment of these three religions as three roads to the same destination.

At roughly the same time that Hinduism was being formalized in the East, Judaism was emerging in the West (or Middle East). Judaism is the first major religion to emphasize a monotheistic, personal God. For Judaism, life is not a burden, but a gift. One of the greatest concerns is ethical living according to the principles of the Torah. Education and particularly study of the Torah, as well as philanthropy are also emphasized.

It is interesting to note that at roughly the same time that Buddha was teaching in India, and Lao Tzu and Confucius were teaching in China, the Hebrew prophets were at their zenith. Their prophetic revelations included the coming of the Messiah (leader or liberator), as well as other events and likely calamities if Jews did not obey God. This breadth of religious creativity in the 6th century b.c. has caused this period to be referred to as a golden age of religions.

For Jews, religion and history are inseparable. Stories in the Jewish tradition include exiles, wandering in the desert, slavery, the Holocaust, and innumerable wars. The latest milestone is the return to form the nation of Israel, which has involved continued strife. This ongoing conflict is considered in detail in Chapter 9.

Christianity resulted from Jesus' protest against Judaism. Much like Buddha, he rejected the castelike system that had evolved in Judaism to differentiate the righteous few from the unhallowed many. The early movement focused on those who were, in effect, disenfranchised by mainstream Judaism.

Christianity revolves around the Jesus story. Telling and celebrating the stories of Jesus' birth, ministry, crucifixion, and resurrection form the backbone of the Christian calendar. The overall story of Jesus—the Christ—who, as the son of God, was sent to earth to redeem all people, has been particularly appealing to those who have felt disinherited by the rest of humankind.

Islam grew out of Mohammed's dislike for the primitive religions of the Arabs and respect for Judaism and Christianity. Mohammed's prophetic revelations led to conflicts with the priests of these religions and eventually armed battles. Mohammed was victorious and returned to Mecca to proclaim it the most holy spot of Islam, the home of the sanctuary of Allah.

Islamic faith rests on the Five Pillars which include faith in Allah, prayer five times each day, almsgiving, keeping the fast of Ramadan (the month leading up to the night when, according to tradition, Mohammed received his first revelations), and a pilgrimage to Mecca at least once in each believer's lifetime. The Koran provides additional, more specific, guidance related to daily life and ethics.

When Mohammed died, he was succeeded by a series of Caliphs. The military orientation of these successors resulted in Islam quickly conquering most of the Middle East, much of northern Africa, and parts of Spain and France. By the 9th century, and continuing for roughly 200 years, Islamic culture flourished in the sciences, arts, philosophy, and poetry. This is often referred to as the Golden Age of Islam.

Two major denominations of Islam have emerged. The Sunnis are those who claim that the first Caliph was Mohammed's true successor. The Shiites, on the other hand, give their allegiance to Mohammed's descendants as the true successors.

Scriptures, Stories, and Myths

It is natural to wonder how the seven religions just discussed differ. To the extent that they are similar, we have a common basis for understanding different cultures. To the extent that they conflict, we can gain insights into issues that should be considered when attempting to innovate in other cultures.

One approach to making such comparisons is to consider the central texts, or scriptures, of these religions and compare the guidance offered relative to various aspects of life. Petersen (1986) makes such a comparison. He compares the scriptures of Hinduism, Buddhism, Taoism, and Christianity and the guidance they offer on love, service to humanity, duty, self-control, and many other topics. He also provides a briefer comparison of Judaism and Islam. His conclusion, which he amply supports, is that all of these religions provide roughly the same guidance about how one should live one's life. While words and metaphors may differ, the messages are essentially the same. Thus, for example, the Golden Rule and the Ten Commandments appear, in one form or another, in all religions.

Petersen, as well as many other authors, has also compared the central stories of many religions. Petersen compares the stories of Krishna, Buddha, and Christ and finds remarkable similarities. He also compares the essential messages of these three central religious figures and finds much in common.

The great religious stories can be called myths, in the most positive sense of the term. Several notable commentators on religion, philosophy, and psychology have discussed myths as follows:

- "Myths are stories of our search for truth, for meaning, for significance" (Campbell, 1988, p. 5).
- ". . . psychologists are coming to realize that myths are myths precisely because they are true. Myths are found in one form or another in culture after

culture, age after age. The reason for their permanence and universality is precisely that they are embodiments of great truths" (Peck, 1987, p. 171).

- "Theology is the timely interpretation of timeless myths. Its most characteristic error is to consider itself timeless" (Kaufman, 1958, p. 224).

Thus, the timeless stories or myths that underlie the world's religions may provide more lasting insights than can be found in the official doctrines and dogmas of these religions. Joseph Campbell's decades of work have provided important contributions to understanding the nature and roles of myths. His many academic treatises are summarized and popularized in his Power of Myth (Campbell, 1988).

He notes that "the folk tale is for entertainment. The myth is for spiritual instruction." Thus, myths play a central role in educating and socializing people. They depict the ideal, as well as the virtues and values associated with this ideal. Campbell argues that God may assume different "masks" in different societies, but the same stories occur everywhere. For example, stories of "virgin births of heroes who die and are resurrected" are revered in many cultures.

His insights, and those of others such as noted earlier, enable understanding the commonality of humankind's search for meaning and hope. However, we often fail to see this commonality because we interpret our own myths as facts. Humans tend to personify, to anthropomorphize, natural forces and try to make the spiritual concrete. As a result, the specifics of our myths or metaphors get in our way. Campbell argues, for instance, that Judaism, Islam, and Christianity are "trapped in their metaphors" and, consequently, fighting it out in the Middle East (see Chapter 9).

If we look at details of sacred texts or themes of the great religious stories, we must conclude that the world's religions have much in common. This leads to the hypothesis that all people have common psychological and social needs and beliefs, which lead all people to find the same stories compelling. This hypothesis is pursued in detail later in this chapter.

First, however, we need to address a more straightforward question. If all of the world's major religions have so much in common, why are there so many religions? Should we expect an eventual convergence and merger, or not?

Sources of Religions

One approach to pursuing these questions is to ask another question, "Why are people religious at all?" While it can be argued that people are incurably religious (Kelsey, 1976), this merely says that they have no choice. Why is that?

It has been suggested that there is a biological basis for religion (Dawkins, 1976; Wilson, 1978). Specifically, it is argued that religious practices have survival benefit. Individuals and communities who are oriented toward such practices tend to prosper. Hence, the genetic makeup of such people is more likely to be propagated.

Religious practices are correlated with prosperity because they reinforce family structure and values, as well as social structure and values. Religion provides people with shared goals, means for coping with problems, and greater stability (Yinger, 1985). This argument is consistent with discussions later in this chapter that focus on the needs and beliefs underlying religious faith.

If we accept, even if we cannot fully explain, the fact of people being inherently religiously oriented, we can move on to consider why there are so many religions, and so many, many sects and denominations within religions. Neibuhr (1929), based in part on the work of Max Weber and Ernst Troeltsch, provides a very insightful analysis.

Neibuhr's analysis focuses on the social sources of religions and denominations within religions. Citing Weber, he notes that Hinduism should be understood in the context of its Brahmin leadership. Buddhism reflects the consequences of monastic supremacy. Confucianism was determined by its relations to the official and literary classes of China. The nature of Islam has been highly influenced by the dominance of the Arabian military class.

Much of Neibuhr's analysis is targeted at Christian denominations. He discusses the impact on early Christianity of Greek thought and the social, religious, political, and economic conditions of the Roman empire. He notes the influence on Roman Catholicism of the Latin spirit and the institutions of the Caesars. Lutheranism was highly affected by the German temperament and political conditions in the church in Germany. Calvinism was greatly influenced by the national character and interests of the economic class to which it appealed, initially in Switzerland.

These observations lead to the conclusion that " . . . the exigencies of church discipline, the demands of national psychology, the effect of social tradition, the influence of cultural heritage, the weight of economic interest play their role in the definition of religious truth" (Neibuhr, p. 17). Due to the very close relationship between religions and the economic, social, and cultural contexts in which they exist, it is quite common for religious institutions to support causes, such as, slavery, that one would think would be inconsistent with their avowed theology and doctrine. Further, conflicts over power, prestige, and possessions are often couched as religious conflicts to gain support. In this way, religions both shape and are shaped by the times and trends.

Despite the high degree of commonality noted earlier, it seems reasonable to conclude that there are strong cultural, social, and economic reasons that make convergence among religions unlikely. Religions are vested in many more interests than just theology and doctrine. They are a primary means for preserving society, not society in general but the particular societies in which they are central.

This leaves the question of why denominations emerge within religions. We have discussed two examples earlier. Buddhism emerged from Hinduism and Christianity emerged from Judaism. In both cases, the leaders of these movements (i.e., Buddha and Jesus) were reacting to the plight of the disenfranchised lower classes.

Neibuhr (1929) explains this phenomenon in terms of the differences between churches and sects. Churches are institutions that one is born into. Membership is a social obligation. Socialization is the primary goal. Churches are usually allied with national, economic, and cultural interests.

Sects are religious movements that one must join, typically in conjunction with some type of religious experience. In fact, religious experience is the primary goal of sects. In contrast to churches, sects are usually separatist in the sense of being primarily minorities outside of the mainstream.

Sects usually form because groups of people are excluded from participating in the mainstream, typically due to their economic, ethnic, or racial status. This exclusion may be direct in that, for example, blacks might not be allowed in white churches. More subtle is indirect exclusion, whereby the theology and doctrines of a church evolve to the point of no longer meeting the needs of the outcast classes.

As a result of exclusion, sects are formed, usually with an agenda more oriented toward social redemption than personal salvation. Characteristics of sects often include much emotional fervor, emphasis on lay leaders, and goals of liberation for the socially disinherited. Associated with these characteristics are usually much discipline and frugality. As a result, sects often provide the basis for much economic progress for its members. As noted in Chapter 2, McClelland (1987) has found enhanced achievement motives reflected in the literature of new religious movements prior to their institutionalism.

Economic progress eventually leads to the sect becoming a church—an institution. As such, the membership becomes more conservative with much emphasis on socializing young people into its traditions. Social redemption no longer necessary, the church now focuses on personal salvation from the excesses possible with the newfound prosperity. Theology and doctrine consequently become more abstract. All in all, this transition results in the once sect, now church, no longer being relevant to whomever comprises the current outcast classes.

The cyclical nature of the above process is described by Neibuhr (1929) as a process where "castes make outcasts and outcasts make castes." He discusses numerous examples of this process. Perhaps the best example is the endorsement of early Christianity by the emperor Constantine in roughly a.d. 300. Christianity thereby became the state church and an institution that had to devote much of its resources to preserving the institution. Some scholars have argued that Constantine's decision was a brilliant way of defusing the constant irritation provided by Christians when they were functioning as a sect.

Kaufman (1958) also discusses the life cycles of religions. Part of this cycle,

Kaufman argues, is a tendency toward the "original sin" of religions—objectifying the divine and accepting as final some dogma. This tendency ignores the necessity of religions being "alive" to change in order to be relevant to the societies they support.

Earlier, a possible biological/genetic basis for religion was noted. We have just discussed cultural, social, and economic reasons for religions. Further, if we accept Neibuhr's cycle, there will always be outcasts seeking to make castes. Therefore, the overall phenomenon of religion is, to a great extent, a predictable outgrowth of biological, psychological, social, economic, and cultural processes and cycles. In the next section, the underpinnings of these phenomena and processes are explored in terms of needs, beliefs, and perceptions.

Summary

In preparation for this exploration, it is useful to summarize what has been covered in this section. The basic chronology of the world's major religions was discussed. These religions were compared in terms of scriptures, stories, and myths. Many similarities were noted, particularly with regard to ethics for everyday living.

Perhaps the greatest theological difference among these seven religions is the emphasis on a personal God in the West, whereas the concept of God in the East is embodied in the sense of a cosmic consciousness (Kelsey, 1976). It has been suggested that the tendency toward more concrete images in the West enabled and perhaps fostered the emergence of the "Age of Enlightenment" which occurred in the 18th century in the West. A parallel period did not occur in the East, in part due to isolationism. However, it has been argued that at least as important an impediment was the East's abstract approach to spirituality.

The Age of Enlightenment led the West to feel that virtually everything can be addressed in a rational, objective manner. In contrast, Eastern cultures do not have this penchant. Thus, for example, my Japanese colleagues have told me that Westerners try to objectively justify all aspects of business relationships. Orientals, on the other hand, accept the subjective nature of business relationships as a normal part of the process. This difference, and its implications, are discussed further in later chapters.

Despite the great similarities among religions from a theological and ethical point of view, economic, social, and cultural forces have resulted in many differences that continue to plague us. These differences lead to disputes and conflicts that are attributed to religious dissimilarities, but should more reasonably be attributed to economic, social, and cultural disparities. Therefore, for example, lower classes inherently see their religions as vehicles for social redemption, while middle classes view their religions as means for personal salvation from their temptations. These two classes are at different points in Neibuhr's cycle, due to economic and social disparities, not necessarily any fundamental theological or ethical differences. Recognition of the true nature of such problems is likely to lead to more innovative approaches to resolving disputes and conflicts.

NEEDS, BELIEFS, AND PERCEPTIONS

The last section focused on the history and, to an extent, the sociology of religion. This material provides necessary background for considering the individual and social psychology associated with needs, beliefs, and perceptions.

This discussion begins by first considering the psychological and social needs underlying religion. The next concern is the nature of beliefs, in particular their source and evolution. A schema for comparing belief systems is discussed. Finally, we return to viability, acceptability, and validity and illustrate the use of the Needs–Beliefs–Perceptions (NBP) Template.

Needs

To begin, it is useful to revisit the central hypothesis underlying this book—people's needs and beliefs strongly affect their perceptions. The relationships among needs and beliefs constitute an important aspect of this hypothesis.

Kaufman (1958) states that "a belief may be accepted because it gratifies us or answers a psychological need." In a similar vein, Casti (1989) says that "there are many alternate realities . . . and any particular brand of reality we select is dictated as much by our psychological needs of the moment as by any sort of rational choice."

Thus, it can be argued that needs strongly influence beliefs. It can also be shown, as later discussions illustrate, that beliefs can dictate needs. Therefore, the relationships among needs and beliefs are complex. The first step in understanding these relationships is consideration of the nature of needs underlying religion.

From a psychoanalytic perspective, it has been argued that piety—tendencies toward devotion and reverence—emerges from infants' helplessness and need for protection by a father figure (Wulff, 1991). It is natural to believe what one is told during formative years and, in general, to believe what other people believe (Peck, 1978). Put simply, children's early years have a tremendous impact on their subsequent beliefs.

Needs that emerge fairly quickly fit roughly into the categories of need for affiliation and need for avoidance in terms of fear of rejection. Beliefs are accepted and endorsed as part of the socialization process (Peck, 1978; Yinger, 1985). Beliefs are part of the socially approved morality which become habits. The result is a common language that facilitates a sense of belonging which, in turn, results in loyalty to a tradition and acceptance of its authority (Kaufman, 1958).

The need to belong, to be part of the group, has a strong effect. Also important is the fear of being rejected by the group. The need to belong results in acceptance of beliefs. The fear of rejection preempts any questioning of these beliefs.

Another avoidance need is the fear of failure. There is a natural tendency to avoid responsibility, particularly for the consequences of one's behavior. Religion helps people to deal with the problem of determining what they are and are not responsible for (Peck, 1978).

People look to their churches, governments, and employers to take responsibility for their welfare. This is especially the case when people perceive that they are unable to take care of themselves. Their concerns can range from being unable to deal with catastrophic health problems—and, therefore, wanting health insurance—to being unable to provide basic food, clothing, and shelter.

Beyond these fundamental needs, people also aspire to fulfill needs to love and create (Kaufman, 1958). Religion has certainly inspired great acts of love (e.g., saintly helping of the downtrodden) and feats of creation (e.g., soaring cathedrals). In this way, religion provides outlets for some of humans' noblest ambitions.

People also need a means for dealing with ultimate concerns. Perhaps the most central of these deals with the "fear and fact of death" (Yinger, 1985). Several avoidance needs can be associated with this issue—avoidance of uncertainty, failure, and the unknown. Religion provides a variety of answers that many people find reassuring.

Other ultimate concerns include guilt and fear in general, yearning for the divine, and fantasies of being divine (Wulff, 1991). People want to avoid consequences, escape the unknown, experience perfection, and perhaps achieve perfection. Religion offers guidelines and support for dealing with these concerns.

Religious faith also provides a way for finding coherence and giving meaning to life (Fowler, 1981). In general, religion provides a means for understanding what life is all about (Peck, 1978). Clearly, people have needs to understand, explain, and attach importance to their lives and their relationships in their families, communities, and cultures.

It would not be at all difficult to map the needs discussed in this section to Maslow's needs (Maslow, 1954) and McClelland's motives (McClelland, 1987), which were identified in the section entitled "Needs" in Chapter 2 of this book. In the discussion of the NBP Template later in this section, such a mapping is considered. Beyond the possibility of this type of organization, it is interesting to examine the extent to which Maslow's hierarchy can be used to predict particular religious beliefs.

MacRae (1977) studied how people's predominant level of need in Maslow's hierarchy affected their notions of heaven and favorite elements of the Bible. He found that there was a relationship. This is not too surprising, considering the frequently observed phenomenon—discussed earlier in this chapter—of lower class people seeing religion as a means for social redemption and middle class people viewing religion as a means for personal salvation. For example, liberation theology appeals to those people who are hoping to find freedom from oppression. Such people find most attractive those portions of the Bible that talk about the poor and meek prospering.

On the other end of Maslow's hierarchy—self-actualization—Wulff (1991) reviews a variety of studies that show self-actualization to be hindered by involvement in traditional religious beliefs and practices, especially if they are conservative. Wulff concludes that while fundamentalists may be sincere and high-minded, their position does not work from a psychological point of view. Campbell (1988) expresses the same concern when he indicates that "anyone brought up in an extremely strict authoritative social situation is unlikely to ever come to the knowledge of himself."

To some extent, it almost goes without saying that people accept, endorse, and promote religion because it meets their needs. Otherwise, why would they? The more interesting question is what specific needs lead to what particular beliefs. Even more interesting is the possibility of inferring needs from espoused beliefs. This would enable use of the characteristics of social, ethnic, and corporate cultures to infer underlying needs, which would provide a competitive advantage when attempting to provide innovative products and services that satisfy needs.

While the discussions so far have begun to expose elements of the knowledge base necessary to accomplishing such inferences, we need a means of structuring what we are uncovering. By considering the sources and evolution of belief systems, the requisite structure starts to emerge.

Beliefs

Why do people believe what they believe? This question motivated much of my research for this chapter. On one level, the answer is simple. The best way to predict somebody's religion is to determine their parents' religion. The best way to predict where people will fall on the liberal–conservative spectrum is to, again, determine their parents' orientation. People are generally born into, and thus often do not choose, their religion.

While this answer is a reasonable explanation for the behaviors of many people, it does not work for everybody. Many people accept, endorse, and promote belief systems that are not totally inherited. Why? Scott Peck (1978) argues that people have a general tendency to make uninformed decisions. They avoid the effort of information seeking and study. They simply accept whatever is the most salient alternative.

This answer provides a useful elaboration of the inheritance explanation. However, there is more. As I pursued the question of why people believe what they believe, it struck me that the question might be better stated as "Why shouldn't people believe what they believe?"

This is a very different question. If people develop a belief system during their formative years, why should they change? Certainly, change has high costs. One must examine the alternatives and risk censure and possibly being ostracized. Unless one's current beliefs are creating major problems, it is much, much easier to just go along, fit in, and not worry about it.

Thus, very few people actually choose their beliefs. They are given a set by their family, community, and culture. Strong motivations are needed to swap this set for a different one. However, the result is not usually a static set of beliefs, passed from generation to generation without change. Typically, people's beliefs evolve as they mature. This section focuses on the nature of this evolution.

The notion that human development proceeds in stages underlies a variety of theories. Jean Piaget's pioneering studies of child development led him to identify four stages of cognitive development (Ginsburg and Opper, 1979):

- Sensorimotor,
- Intuitive,
- Concrete operational, and
- Formal operational.

These stages correspond with relatively primitive coordination of perception and action, subsequent integration of language and thought, initial abilities for causal reasoning, and eventual abilities for formal reasoning, respectively.

Lawrence Kohlberg's (1981) studies of the development of moral standards led him to describe this process in terms of three stages:

- Preconventional,
- Conventional, and
- Principled.

In the preconventional stage, standards are based on what will avoid punishment or bring pleasure. The conventional stage is characterized by acceptance of standards that will earn the approval of others. In the principled stage, standards are viewed in terms of obligations to others and one's own conscience.

Erik Erikson (1982) has posited the eight stages of human development shown in Figure 3.2. Each stage is characterized by a central conflict that humans must face as they develop. Erikson argues that these conflicts must be dealt with at physiological, psychological, and social levels.

For some people, the conflicts in Erikson's framework are not resolved in a manner that enables growth. For example, they may become mired in isolation, stagnation, or despair. Psychiatrist Scott Peck (1978) suggests, somewhat tongue-in-cheek, that about 1 in 20 grown-ups succeeds in becoming adult.

STAGE	CENTRAL CONFLICT
Infancy	Basic Trust vs. Mistrust
Early Childhood	Autonomy vs. Shame, Doubt
Play Age	Initiative vs. Guilt
School Age	Industry vs. Inferiority
Adolescence	Identity vs. Identity Confusion
Young Adulthood	Intimacy vs. Isolation
Adulthood	Generativity vs. Stagnation
Old Age	Ego Integrity vs. Despair

FIGURE 3.2. Eight stages of human development (adapted from Erikson, 1982).

James Fowler (1981) integrates the theories of Piaget, Kohlberg, and Erikson to synthesize a framework for understanding humans' development of belief systems. His studies of what he terms faith development has led to the identification of the six stages of faith shown in Figure 3.3.

In the intuitive-projective stage, children's imagination and their tendencies to imitate combine to yield fantasylike images associated with beliefs. With the emergence of concrete operational thinking, children systematically accept and literally interpret the beliefs, observances, moral rules, and attitudes of the traditions in which they are being raised. The development of formal operational thinking enables a transition from the mythic-literal stage to the synthetic-conventional stage, which is characterized as the conformist stage where others' expectations and judgments are central influences.

Many adults spend their whole lives at the synthetic-conventional stage. For others, conventional beliefs are undermined by experiences that promote reflection on the origins and nature of beliefs. The resulting individuative-reflective stage is characterized by a realization of the relativity of one's inherited world view. Reliance on external approval and authority is discarded. One's own abilities to reflect and reason become central.

Transition to the conjunctive stage is characterized by a recognition of the paradoxes and perhaps contradictions in life. This leads to an openness to the

STAGE	MEANING
Intuitive-Projective	Fantasy-filled imitation
Mythic-Literal	Literal interpretation of beliefs, observances, moral rules, and attitudes
Synthetic-Conventional	Conformity tuned to expectations and judgments of others, particularly traditional authority roles
Individuative-Reflective	Discarding of many myths with emphasis on conscious mind and critical thought
Conjunctive	Symbolic power reunited with conceptual meanings and recognition of the role of the unconscious
Universalizing	Loving and sharing inclusiveness across all social, cultural, ethnic, and economic distinctions

FIGURE 3.3. Six stages of faith (adapted from Fowler, 1981).

traditions of other communities, as well as a realization that ultimate truth cannot be exposed solely by reflection and reasoning. Consequently, symbolic power is reunited with conceptual meanings. Symbolic experiences again have power even though their relativity and ultimate inadequacy are understood.

The universalizing stage is characterized by those rare individuals who develop to the point of embracing the diversity of the world in a loving and sharing manner. Selflessness leads them to a strong feeling of inclusiveness across all of humanity's usual distinctions. Fowler (1981) suggests Gandhi, Mother Teresa, and Martin Luther King Jr. as examples of people who have attained the universalizing stage.

The notion of stages of faith or belief appears to be widely applicable. McClelland (1987) asserts that parallels of stages of faith can be found in all of the world's major religions. Kaufman (1958) suggests that people of different religions at similar stages of faith are more alike than those within the same religion at different stages of faith. Hence, for example, it could be argued that fundamentalist Christians have more in common with fundamentalist Muslims, at least in terms of the nature of their beliefs, than they do with liberal Christians. Of course, as Campbell (1988) would argue, they would have difficulty recognizing this because their metaphors would get in the way.

To conclude this discussion of the psychology of beliefs, it is useful to consider Wulff's schema for depicting different perspectives on the psychology of religion (Wulff, 1991). This schema, shown in Figure 3.4, has two axes. The horizontal axis distinguishes literal from symbolic perspectives. The vertical axis differentiates perspectives that include the possibility of a transcendent reality from those that limit reality to that of the concrete world.

In the lower left, literal disaffirmation is typified by sociobiology and behaviorism which attempt to explain all human behaviors, religious or otherwise, in terms of biological and physical processes, without resorting to any concepts such as mind, ego, and so on. The lower right represents a perspective that remains in the nontranscendent world, but emphasizes the role of the symbolic in terms of frameworks such as psychoanalysis and ego psychology.

The upper-right perspective focuses on the symbolic nature of beliefs and includes the possibility of a transcendent reality typified by Fowler's concept of conjunctive faith. Finally, the upper-left perspective is exemplified by the literalist emphasis of religious fundamentalism.

Inclusion of Transcendence

LITERAL AFFIRMATION (Religious Fundamentalism)

RESTORATIVE INTERPRETATION (Conjunctive Faith —Fowler)

Literal — Symbolic

LITERAL DISAFFIRMATION (Sociobiology, Behaviorism)

REDUCTIVE INTERPRETATION (Psychoanalysis, Ego Psychology—Erikson)

Exclusion of Transcendence

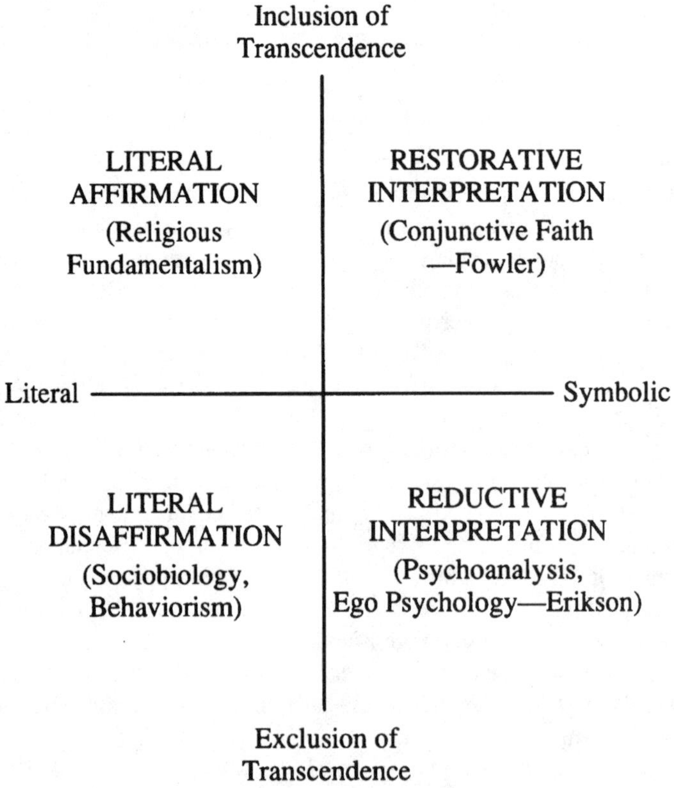

FIGURE 3.4. Perspectives on the psychology of religion (adapted from Wulff, 1991).

Perceptions

The last two sections, on needs and beliefs, covered quite a bit of ground. It is easy to lose track of why the history, sociology, and psychology of religion are relevant to the problems and issues addressed by this book. A summary and integration are needed.

Figure 3.5 provides a summary of the needs identified in the discussions of culture and religion in this chapter. I hasten to note that this compilation is augmented at the conclusion of several chapters in this book. Thus, this list should be viewed as a "prototype" compilation.

In preparing this list, I tried to order it roughly—very roughly—according to Maslow's hierarchy (Maslow, 1954). Therefore, the first set deals with basic physiological and safety needs. The second set of needs deals with belongingness and fears of not being accepted. The third set is concerned with the benefits of belonging, a portion of which may involve self esteem.

The fourth set emphasizes needs associated with thoughtful and perhaps reflective dealings with the world. My guess is that these types of need would, for example, characterize those at Fowler's individuative-reflective stage of faith. These types of need, as well as others, emerge again in the discussions of science and technology in Chapter 4.

The fifth set of needs relates to desires to deal with ultimate issues in life. It differs from the fourth set in that the concerns are truth and significance, not just consistency and coherence. Both sets are similar, however, in that they deal with aspects of self-actualization.

While this list roughly matches Maslow's hierarchy, it is also easy to see McClelland's motives (McCllelland, 1987) in Figure 3.5. The first, second, and third sets have many elements of needs for affiliation and needs for avoidance. The fourth and fifth sets, somewhat less clearly, hint at elements of needs for achievement and needs for power.

Figure 3.6 provides a summary of the beliefs identified in the discussions in this chapter. This compilation is more heavily religiously oriented than the sets of needs in Figure 3.5. As this compilation of beliefs is augmented via discussions in later chapters, a wide variety of other types of belief are added.

As it stands, the list in Figure 3.6 includes some vary basic elements of beliefs. An interesting contrast is between a literal, objective, personal God and a symbolic, subjective, cosmic consciousness. This very rough characterization of Western vs. Eastern perspectives provides a few insights in discussions in later chapters.

While Figures 3.5 and 3.6 focus on individual needs and beliefs, respectively, Figure 3.7 emphasizes what can be called societal needs. To avoid chaos and anarchy, society needs order and a means of instilling the value of order in its members. Religion can provide this means. However, if religion is to be accepted,

- Need to deal with fear of helplessness
- Need for protection
- Need to deal with fear of uncertainity
- Need to deal with fear of the unknown
- Need to deal with fear of death

- Need for acceptance
- Need for approval
- Need to belong
- Need to be judged well by others
- Need to fit into tradition
- Need to imitate
- Need to deal with fear of rejection
- Need to deal with guilt
- Need to deal with fear of failure and consequences

- Need to love
- Need for shared goals
- Need to fulfill obligations to others
- Need for social redemption

- Need for stability
- Need for coherence
- Need to be in control
- Need for autonomy
- Need to be consistent
- Need to reason things out
- Need to deal with contradictions
- Need to deal with paradoxes

- Need to understand life
- Need for truth
- Need for meaning
- Need to create
- Need for significance
- Need to relate to the divine
- Need for personal salvation

FIGURE 3.5. Summary of individual needs identified.

- Belief in being chosen
- Belief in being an outcast
- Belief in one's own abilities
- Belief in literal interpretation
- Belief in symbolic interpretation
- Belief in importance of objectivity
- Belief in value of subjectivity
- Belief in transcendent reality
- Belief in a personal God
- Belief in a cosmic consciousness
- Belief in value of multiple perspectives

FIGURE 3.6. Summary of individual beliefs indentified.

it must adapt its dictates to the societal context in which it operates. On occasion, this requires that religion endorse practices that are inconsistent with its avowed doctrine. Without such endorsements, religion risks losing its power to meet societal needs.

A few other observations in this chapter should be reviewed. As Niebuhr (1929) said, "Castes make outcasts and outcasts make castes." Outcasts are created directly by prohibiting membership in the "in group," or they are created indirectly by theology and doctrine that are irrelevant to the "out group." In either case, those who are economically, socially, culturally, or racially outcast form their own sects, which eventually become churches. These sects attempt to address grievances that are mainly economic, social, cultural, and racial. Conflicts often result. These conflicts are called religious conflicts, but they usually have relatively little to do with religion.

The information in Figures 3.5–3.7 represents the beginning of a knowledge base with which we address the four archetypical innovation problems introduced earlier and discussed at length in Chapters 6–9. In the remainder of this section, an example is considered to illustrate how this evolving knowledge base can be used.

For this example, I have chosen a recent experience. Last year, I ran a workshop for would-be entrepreneurs in South Africa. A foundation had made venture capital available, with very favorable terms, for the purpose of job creation in the country's beleaguered economy. I was asked to provide an overview lecture on technological innovation and business formation. I then served as the facilitator of

- Need for social order
- Need for harmonious relationships
- Need to socialize young people
- Need to foster principles for ethical living
- Need to communicate via scriptures, stories, myths, ceremonies, and rituals
- Need to adapt to cultural, social, and economic interests
- Need to support causes even if they are inconsistent with theology
- Need to provide means for survival of society

FIGURE 3.7. Summary of societal needs indentified.

four working groups. These groups, of roughly 20 people each, focused on the concerns underlying people's decision to start a new venture or not.

Most people initially expressed the belief that they had the technological know-how and all they needed was the capital. They further believed that capital was either unavailable, or was only available with unacceptable terms. However, the foundation's announcement that day had just made clear that capital was now available with surprisingly attractive terms.

One would think that people would now jump at the chance. They didn't. We spent much of the time in the working groups trying to understand their continuing reluctance. Why was the package of technological know-how and easily available capital not enough for most people?

Consider this problem in terms of validity, acceptability, and viability. Clearly, most people perceived the validity of this package negatively—it did not solve their problem. After much discussion, it emerged that acceptability was viewed negatively because starting their own ventures would require too much change. Finally, viability was perceived negatively because likely benefits appeared to be dwarfed by costs.

What beliefs were underlying these perceptions? Considerable discussion led to the conclusion that while most people believed in their technological know-how, they did not believe in their business know-how. Further, many thought that going into business would dramatically change their lifestyles, with much stress and little time for families. Finally, they believed that the primary benefit of starting a business was to give themselves jobs, so they would not have to worry about unemployment. However, most of them already had jobs. While they were worried about the security of those jobs, they nevertheless had jobs at the moment.

These beliefs are compiled in the example NBP Template shown in Figure 3.8. Once these beliefs were identified, the next question concerned what needs were associated with these beliefs. The needs column in Figure 3.8 indicates the answers that emerged.

People had much concern about uncertainty and unknowns. They were not really sure what it took to start and run a business. They also were not sure of the impact on their lifestyles. They were not negative about these things per se; they were negative because they could not cope with the apparent levels of uncertainty and extent of the unknowns. Once these problems were understood, the foundation committed to providing training as well as business planning tools and advice at

ATTRIBUTE	PERCEPTIONS	BELIEFS	NEEDS
Viability	Positive		
	Negative	Business primarily for purpose of having a job	Lack of a need to create and achieve
Acceptability	Positive		
	Negative	Lifestyle change potentially very large	Need to deal with fears of uncertainty and unknowns, need for stability and control
Validity	Positive		
	Negative	Business know-how insufficient	Need to deal with fears of uncertainty, unknowns, failure and consequences, need to be judged well by others

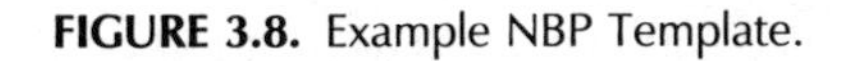

FIGURE 3.8. Example NBP Template.

a new venture support center. Many would-be entrepreneurs responded quite positively.

The problem underlying benefits not being perceived as greater than costs was best explained by lack of a need! Many people were primarily concerned with having a secure job and maintaining their lifestyles. While they realized that fundamental changes occurring in South Africa would inevitably affect them, their jobs, and their lifestyles, not much had happened yet. Perhaps they could just hold on?

At one point I asked, "Doesn't anyone here want to start a business just to see if they can do it?" No one responded. There appeared to be a lack of a need to create and achieve. According to McClelland's analyses (McClelland, 1987), this is a very fundamental problem.

This conclusion could easily have led to a dramatic lowering of expectations of what this group of would-be entrepreneurs could accomplish. Fortunately, further discussion led to an important insight. Several people eventually said that they had a repressed need to create and achieve, but they were unwilling to vocalize these needs—admit them to themselves—because they were so uncertain about their own abilities. They felt that with appropriate training and ongoing support they could eventually gain enough confidence to take business risks simply because they wanted to achieve.

Thus, we see an interesting interaction. Viability was perceived as negative because people were unwilling to admit its positive side due to concerns about acceptability and validity. The NBP Template helps to uncover subtleties such as this. Numerous examples in later chapters provide insights into other subtleties.

Summary

This section has provided an initial in-depth look at needs, beliefs, and perceptions. The history, sociology, and psychology of religion, as flavored by culture, formed the basis for this exploration. Using the resulting knowledge base as a starting point, we are in position to explore other contexts to elaborate our understanding of needs, beliefs, and perceptions.

ORGANIZATIONAL CULTURES AND RELIGIONS

Thus far in this chapter, religions have been considered in the "official" sense of the term. However, this is not a book about religions per se. This material was reviewed to provide the basis for initial elaboration of the Needs–Beliefs–Perceptions Model.

In this section, the discussion broadens to consider organizations other than those normally associated with religions. We first look at organizational denominations—forms of organization characterized by particular sets of needs and beliefs. Next, needs, beliefs, and perceptions in organizational settings are discussed. Finally, the implications of these conclusions are considered relative to the four archetypical innovation problems.

To some extent, this section is a preview of the types of discussion in Chapters 6–9. It is necessarily a cursory preview, since in-depth treatment is delayed until the later chapters. Nevertheless, the following discussion provides a glimpse of how consideration of the relationships among needs, beliefs, and perceptions can provide important insights.

Organizational Denominations

When we think about types of organizations or enterprises, it is typical to consider distinctions such as industry vs. government vs. academia. This distinction reflects, for example, the differences of making a profit for shareholders, serving constituencies' interests, and pursuing knowledge and education. These differences mainly reflect the contexts in which these types of enterprises operate.

Enterprises can also be classified in terms of how they are organized to operate. Corporations, partnerships, and proprietorships are common forms of organization. Joint ventures and consortia are increasingly common. These forms of organization specify an enterprise from a legal point of view.

The nature of an organization can also be considered in terms of how its structure enables its operations. Rosabeth Moss Kanter (1989) contrasts the traditional hierarchy with the typical entrepreneurial form of organization. The traditional hierarchy has clear reporting relationships, policies for getting plans and activities approved, and well-understood "ladders" for furthering one's career. The entrepreneurial enterprise, on the other hand, has few if any levels of management, creates policies on an as-needed basis, and provides opportunities for achievement rather than promotions.

The hierarchical organization tends to be stable and predictable. Its tendency to become bureaucratic makes it difficult to change and, therefore, it has a difficult time innovating. The entrepreneurial organization, in contrast, tends to be marginally stable, constantly changing, and *potentially* very innovative. If marginal stability becomes instability, entrepreneurial enterprises may not survive and potential innovations will not be realized.

Kanter argues for a form of organization that she terms "post entrepreneurial." This new form is characterized by its contrast with bureaucratic forms of organization, as shown in Figure 3.9. While the traditional organization focuses on positions and the responsibilities and perquisites of positions, the postentrepreneurial organization emphasizes people. Rather than focusing on following the rules and using formal structures, the post-entrepreneurial enterprise emphasizes creativity and results via communication.

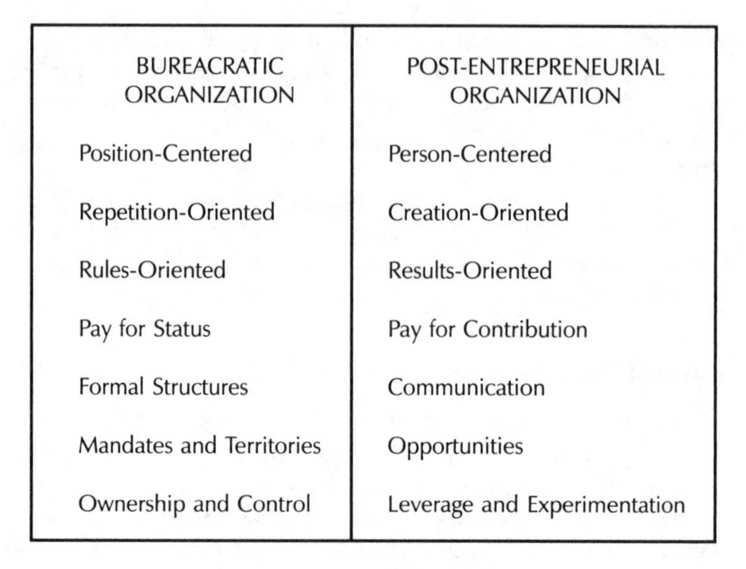

BUREACRATIC ORGANIZATION	POST-ENTREPRENEURIAL ORGANIZATION
Position-Centered	Person-Centered
Repetition-Oriented	Creation-Oriented
Rules-Oriented	Results-Oriented
Pay for Status	Pay for Contribution
Formal Structures	Communication
Mandates and Territories	Opportunities
Ownership and Control	Leverage and Experimentation

FIGURE 3.9. Contrasting types of organization (adapted from Kanter, 1989).

In the bureaucratic organization, people are paid for their position—their status—and their ownership and control of territories or turf. Post-entrepreneurial organizations pay people for their contributions, which requires that they seek opportunities, gain leverage, and experiment. People are rewarded for doing things and making things happen rather than directing and monitoring the performance of others. Team leaders are emphasized over team managers.

The post-entrepreneurial form of organization has emerged as entrepreneurial enterprises have "grown up" or as traditional enterprises have gone through major changes, often in response to crises. This new form of organization has led to many changes such as "delayering" and "dislocated" professionals. Put more plainly, many companies have laid off substantial fractions of their middle managers.

This trend has challenged many strongly held beliefs. In Atlanta, a city where virtually every Fortune 500 company has its southeastern United States headquarters, I have talked with numerous people in the following situation. They graduated from college about 10 to 12 years ago, with a degree in liberal arts or social sciences. They took a job with a Fortune 500 company and had worked themselves up a few rungs on the corporate ladder to a middle management position. Virtually all of their expertise, in their opinions at least, is in their knowledge of how the company works and their skills in getting things done within the tangled hierarchy of the corporation.

These people believed that they would slowly but surely make it up the ladder and be with their company for all of their careers. They moved when told—being in Atlanta for a few years was, in effect, "getting your ticket punched" to be eligible for higher positions. They might change companies, but only on their terms if they were "raided" by competitors.

When in recent years such people were "dislocated" as a result of "delayering," they were shocked. Their beliefs in how corporate America works were violated. Their needs for achievement, power, or affiliation suddenly took a backseat to their avoidance needs. They feared being perceived as failures and possible financial ruin since their commitments had been predicated on beliefs of continual growth of earnings in a "safe" job. Their company, and most others to which they could apply for jobs, had suddenly changed religions. The new theology and doctrines did not require the functions that such people had come to pride themselves in being able to perform skillfully.

Needs, Beliefs, and Perceptions

In this section, the needs, beliefs, and perceptions underlying phenomena such as just discussed are explored. This exploration first involves consideration of needs related to organizational membership, the nature of beliefs within organizations, and examples of differences across cultures. The impact of organization changes such as illustrated in the above example is then considered in terms of its effect on needs and beliefs.

Membership in an organization usually involves much more than just practical ends. Beyond providing a means to earn a living, for example, organizations also provide a social context for members of the organization. Senge (1990), Thurow (1992), and Wilkins (1989) have suggested a variety of needs that people seek to have met in the organizations in which they are members—see Figure 3.10.

- Need to belong, be part of a group
- Need to be part of a worthwhile group
- Need to be secure in their membership
- Need to be held in esteem
- Need to feel useful
- Need to believe that they can make a difference
- Need for continuity
- Need for creativity

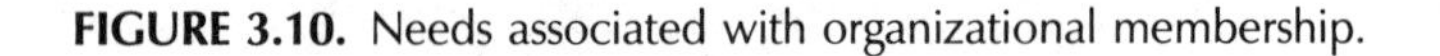

FIGURE 3.10. Needs associated with organizational membership.

If the organization does not provide the means to meet these needs, people's performance may suffer. For instance, if people feel powerless within the organization—that they are not useful and cannot make a difference—they are unlikely to undertake efforts that, in fact, may be of great value to the organization. They may not even make simple suggestions for improvements.

Another need common to most people is a desire for continuity. Most people want the future to be very similar to the past. On the other hand, there is also often a desire for creativity which usually involves change. A fundamental tension results (Senge, 1990). Balancing the tension between continuity and creativity is a basic issue in enabling the organization—this issue is discussed in more detail in Chapter 7.

Alan Wilkins (1989) has considered the nature of people's beliefs in organizations using a construct he calls motivational faith. The four elements of motivational faith are listed in Figure 3.11.

Wilkins argues that people's willingness to share the organization's vision of its mission depends to a great extent on their beliefs about themselves and others being treated fairly, the realistic possibilities for pursuing the vision, and their own abilities to contribute to this pursuit and be valued for contributing.

As discussed in Chapter 2, the nature of these beliefs can be described in terms of people's mental models of the organization (Senge, 1990). People's mental models of how the organization functions and their roles within it are central to their beliefs about the organization. In addition, their assumptions about what is important, what can be accomplished, what problems are inevitable, and so on frame their perceptions of whether the organizational aspirations are reasonable, and what they individually can contribute. These types of assumptions are often culturally determined by the organization's history and stories about this history. Stories have been found to be an important means for communicating an organization's culture (Senge, 1990; Wilkins, 1989).

Stories can capture and operationalize ongoing truths that are formalized in an organization's mission statement. Stories, and symbols associated with these stories, can embody and communicate the nuances of the culture.

- Belief in the fairness of the organization
- Belief in the fairness of the people in the organization
- Belief in the abilities of the organization
- Belief in one's own abilities

FIGURE 3.11. Beliefs associated with organizational membership (adapted from Wilkins, 1989).

Success stories show how the organization has dealt with adversity and crisis. They also provide tales of quests that can be celebrated. I have found that participants in my frequent workshops are particularly intrigued with failure stories. Of special interest are stories of decisions that seemed intuitively correct, but turned out to be bad decisions because, for example, crucial factors were ignored. Discussions of how these factors came to be ignored often provide insights into organizational stumbling blocks that, once identified, are easily avoided. Stories, in this way, teach in a manner that is much more compelling than traditional didactic teaching. The set of stories that underlie an organization's culture becomes, in effect, the organization's scriptures.

Needs and beliefs associated with organizations can vary across cultures. Lester Thurow (1992) illustrates such differences in his comparison of business practices in the United States, Japan, and Europe. He stresses in particular the emphasis on the individual in the United States vs. emphasis on the team in Japan and, to an extent, Europe. In the United States, success is usually attributed to individual skill and prowess —outstanding individuals are installed in one or more of a wide variety of halls of fame. In contrast, in Japan in particular, but also Europe, success is attributed to teamwork.

Thurow generalizes this contrast by comparing what he calls individualistic capitalism and communitarian capitalism. The beliefs of these two denominations of capitalism are summarized in Figure 3.12. Beliefs associated with individualistic capitalism reflect the mythical "cowboy" of American culture, taking problems in his hands and solving them, despite the group and its inability to perform. Communitarians, on the other hand, believe in cooperation, responsibility to society, the value of teamwork, and loyalty to the team and firm.

INDIVIDUALISTIC CAPITALISM	COMMUNITARIAN CAPITALISM
Individual Values	Business Groups
Entrepreneurs	Social Responsibility
Individual Responsibility	Teamwork
Profit Maximization	Firm Loyalty
Consumer Economics	Producer Economics

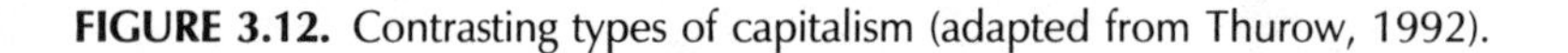

FIGURE 3.12. Contrasting types of capitalism (adapted from Thurow, 1992).

The differences between consumer and producer economics provide additional important distinctions between these two denominations. Consumer economics focuses on making money in order to enjoy the things that this money can buy—the emphasis is on working to consume. Producer economics focuses on making enough money to be able to improve abilities to produce—emphasis is on generating capital that can be reinvested in order to become better at generating capital. As a consequence of this difference, Thurow argues that the United States is concerned with making money as an end in itself, while Japan, for instance, is concerned with job creation and community building.

Geert Hofstede (1980) discusses this distinction among business cultures, as well as several other distinctions. His extensive questionnaires (e.g., 116,000 in all) involved asking people in 40 countries 150 questions, including 60 about beliefs and values. The resulting data were subjected to a variety of statistical analyses. Based on these analyses, four attributes of business cultures were identified:

- Power distance,
- Uncertainty avoidance,
- Individualism, and
- Masculinity.

Power distance reflects the extent to which a society accepts inequalities and power hierarchies as natural phenomena whereby a few are in charge and all others are subordinate. Uncertainty avoidance is concerned with the extent to which a society feels threatened by change and will not tolerate deviant ideas and behaviors. Individualism reflects the extent to which society is viewed as a loose-knit social structure, where people take care of themselves, vs. collectivism, where members of the social group are expected to look after each other. Finally, masculinity concerns the degree to which assertiveness and the acquisition of money and things are extolled, vs. femininity, where caring for others, quality of life, and people are the primary concerns.

Using several cluster diagrams, Hofstede (1980) depicts relationships among the countries studied. While the results are certainly not crisp, frequently appearing clusters include

- Developing countries,
- Scandinavian countries,
- Germanic countries, and
- English-speaking countries.

The United States, obviously, is included in the last of these clusters.

The implications are that United States business practices are likely to be quite compatible with those in countries in the same cluster as the United States. This agrees with my experience in Europe, Asia, Africa, and South America. In contrast, when doing business with countries not in the same cluster as the United States, cultural differences may play a bigger role. For example, the developing countries score much higher than the United States on power distance and uncertainty avoidance, and much lower on individualism. Consequently, based on Hofstede's analysis, it would seem reasonable to expect potential business partners in developing countries to be quite willing for remote powers to be "in charge" in order to maintain stability, and also to be more concerned with collective accomplishments (i.e., job creation) rather than individual accomplishments. Business deals that emphasize United States values may not be of great interest to potential partners without the same values.

Thus, it is easy to see that there are substantial cultural differences. While, as we discussed extensively earlier, the world's religions espouse very similar ethical tenants, the ways in which different cultures deal with the world differ substantially, depending on the specific situations different groups have encountered and the historical evolution of groups' experiences. As a consequence, below the surface, disputes attributed to religious differences are not so much due to religious differences per se as they are due to cultural, social, and economic differences. This conclusion provides the basis for creative approaches to resolving sociotechnical disputes (Chapter 8) and political conflicts (Chapter 9).

The impact of needs and beliefs is perhaps greatest when an organization tries to change, perhaps because of a crisis that necessitates change. Kanter (1989) discusses the substantial changes that United States corporations have gone through, and will continue to go through, in order to compete in the rapidly evolving global economy. An excellent example of such changes is the transition from "pay for position" to "pay for performance." In other words, gaining a succession of positions—moving up the ladder—no longer guarantees that salary and perquisites will continually increase. What matters more, and will increasingly be emphasized, is one's contribution to the enterprise's value added in the marketplace.

The dramatic shift in emphasis, which has only begun to emerge, results in the types of phenomena discussed earlier. Once the value of middle managers was questioned, the inquiry quickly evolved to the point of questioning whether or not their positions, and hence their paychecks, were justified in terms of value added. In many cases, as the onslaught of layoffs indicate, the answer was that these types of positions did not result in valuable performance. Often, the layoffs were not due to a lack of money to pay people. They were due to a lack of value being placed on what such people did.

The impact of such layoffs can have a tremendous impact on those who are not laid off. Kanter suggests that the result can be a "crisis of commitment" that yields a need to reaffirm organizational membership. People may no longer be sure that they can count on the organization. They also may not be sure that they want to remain, given that major changes are necessary.

Kanter (1989) and Wilkins (1989) argue that major changes should be accompanied with opportunities for those that remain to mourn the past that is gone and grieve over losses. In other words, management should not deny the significance of the change and make light of what is being discarded. Stories and symbols should be used to recall and revere those aspects of the past that deserve such status. The changes at hand can, for instance, be celebrated as a passage to a new era of the organization's development. In this way, the past and the future are linked, which results in people's needs for continuity and tradition being met.

I do not mean for such transitions to sound easier than they really are. As Senge (1990) notes, new insights of opportunities—or necessities—for change may not be recognized or pursued if they conflict with deeply held images. I have been involved with numerous enterprises, including my own, that have attempted transitions from primarily research and development organizations to broader enterprises, providing products and services as well as research results. Much of people's reluctance to go along with such changes, even though market forces dictated these changes, could be attributed to their self-images of being "researchers," not practitioners. They did not want to give up their lab coats for business suits. However, they were not being asked to do so. They were still being asked to be researchers. But, they did not want the organization to shift its values such that other goals, in addition to research, were of equal importance. If this were to happen, they perceived that their importance would decrease.

Note that people did not explain their reluctance in this way. They usually had much more practical explanations for why necessary changes should be avoided. However, by reflecting on their behaviors in terms of needs, beliefs, and perceptions, it was possible to identify more plausible causes. This process of identification is illustrated in detail in Chapters 6–9.

Of particular importance, of course, is what one can do with such insights. Wilkins (1989) discusses this issue and concludes that organizational "prophets" have to listen to criticisms and affirmations and adapt accordingly. Thus, for the example of an R&D organization having to become more market oriented, the answer to criticisms was not to try to convince people of the necessity of the change. Instead, the response was to extol the importance of R&D in enabling the new future, as well as its central role in continuing to define the future. In other words, the message was that R&D would be more important, not less important, because of the changes at hand.

Dealing with problems such as this in a positive way is a characteristic of what Tom Lloyd (1990) calls "nice" companies. He argues that the Darwinian "survival of the fittest" model of management is disappearing. While he does not state his conclusions in these terms, he is, in effect, arguing for the importance of understanding the needs and beliefs that underlie people's perceptions of the organization, its values, and its agenda. Lloyd, as well as Kanter, Senge, Thurow, and Wilkins, suggests that scriptures, stories, and myths play a much larger role in organizational cultures and religions than we might expect.

Implications

What insights are gained by characterizing organizations in terms of cultural and religious attributes? First of all, significant leverage is gained by realizing the strong influence of cultural and social phenomena regardless of whether the context is religion, business, or the community. Needs and beliefs influence virtually everything we do.

From the perspective of organizational change, it is important to realize that change is not just inconvenient. It can challenge basic assumptions and long-held beliefs. From this perspective, it is not surprising that crises may be required to motivate real changes in world views (Thurow, 1992). In fact, it may be necessary to allow crises to emerge to get people to take the need to change seriously.

Lloyd (1990) argues that true strategic change can only be accomplished if managers' aspirations, attitudes, and belief systems change. Similarly, Thurow asserts that beliefs about the right management style change very slowly and only if forced. Consequently, it is not just the needs and beliefs of the workers and employees in general that must be understood. Top management also has needs and belief systems that are just as difficult to change.

It is interesting to come full circle. How do formal religions and organizational religions relate? Yinger (1985) and Mokyr (1991) consider this question and reach similar conclusions. Traditional religions tend to inhibit, or at least not support, economic development. As noted in earlier discussions, institutionalized churches emphasize maintenance of the status quo in terms of mainstream values and priorities. Such an emphasis is not supportive of entrepreneurship that involves questioning of values and priorities.

To summarize this section, organizational value systems and beliefs can be conceptualized as cultures and religions. While the resulting constructs are not identical to religions in the usual sense of this term, the phenomena underlying these social systems are similar. People's needs and beliefs interact to influence perceptions that determine the difficulties, or ease, with which organizations deal with change. The nature of these relationships is explored in depth in Chapter 7.

SUMMARY

This chapter has covered an enormous amount of ground. The history, sociology, and psychology of the world's religions were reviewed and contrasted. This led to initial elaboration of the NBP Template and a brief illustration of its use. The notion of organizational cultures and religions was introduced to provide the basis for subsequent elaboration in Chapter 7.

This chapter also served to provide some "meat" to the skeletal NBP Model outlined in Chapter 2. Within the rich and contextually laden context of culture and religion, the model seems to be plausible and make sense. To provide convergent evidence of the plausibility of the model, the next chapter pursues a completely different perspective. By immersing ourselves in the context of science and technology, we can evaluate the roles of needs and beliefs in a context where these terms are seldom used.

REFERENCES

Campbell, J. (1988). *The power of myth.* New York: Doubleday.

Casti, J. L. (1989). *Paradigms lost: Images of man in the mirror of science.* New York: Morrow.

Dawkins, R. (1976). *The selfish gene.* Oxford: Oxford University Press.

Erikson, E. H. (1982). *The life cycle completed: A review.* New York: Norton.

Fowler, J. W. (1981). *Stages of faith: The psychology of human development and the quest for meaning.* New York: Harper and Row.

Ginsburg, H., and Opper, S. (1979). *Piaget's theory of intellectual development.* Englewood Cliffs, NJ: Prentice-Hall.

Hofstede, G. (1980). Motivation, leadership, and organization: Do American theories apply abroad? *Organizational Dynamics*, 42–63.

Kanter, R. M. (1989). *When giants learn to dance.* New York: Simon and Schuster.

Kaufman, W. (1958). *Critique of religion and philosophy.* Princeton, NJ: Princeton University Press.

Kelsey, M. T. (1976). *The other side of silence: A guide to Christian meditation.* New York: Paulist Press.

Kohlberg, L. (1981). *The philosophy of moral development: Moral stages and the idea of justice.* San Francisco: Harper and Row.

Lloyd, T. (1990). *The 'nice' company: Why 'nice' companies make more profits.* London: Bloomsbury.

MacRae, D. R. (1977). The relationship of psychological needs to God concept and religious perceptions. *Dissertation Abstracts International, 38*, 1954B–1955B.

Maslow, A. H. (1954). *Motivation and personality.* New York: Harper.

McClelland, D. C. (1987). *Human motivation.* Cambridge, UK: Cambridge University Press.

Mokyr, J. (1991). *The lever of riches: Technological creativity and economic progress.* Oxford, UK: Oxford University Press.

Neibuhr, H. R. (1929). *The social sources of denominationalism*. New York: Henry Holt.

Peck, M. S. (1978). *The road less traveled: A new psychology of love, traditional values and spiritual growth*. New York: Simon and Schuster.

Peck, M. S. (1987). *The different drum: Community-making and peace*. New York: Simon and Schuster.

Petersen, R. (1986). *Everyone is right: A new look at comparative religion and its relation to science*. Marina del Rey, CA: DeVorss.

Senge, P. M. (1990). *The fifth discipline: The art and practice of the learning organization*. New York: Doubleday/Currency.

Thurow, L. (1992). *Head to head: The coming economic battle among Japan, Europe, and America*. New York: Morrow.

Wilkins, A. L. (1989). *Developing a corporate character: How to successfully change an organization without destroying it*. San Francisco: Jossey-Bass.

Wilson, E. O. (1978). *On human nature*. Cambridge, MA: Harvard University Press.

Wulff, D. M. (1991). *Psychology of religion: Classic and contemporary views*. New York: Wiley.

Yinger, J. M. (1985). Social aspects of religion. *Encyclopedia Britannica, 26*, 538–547.

Chapter **4**

Science and Technology

It is quite natural to pursue an understanding of needs, beliefs, and perceptions by studying the nature and evolution of culture and religion. We tend to think that the compartment of our lives labeled "Culture and Religion" includes things like needs and beliefs. However, we don't usually associate the other compartments in our lives with these concepts.

In the last chapter, we began the process of exploring other compartments by discussing organizational settings. In this chapter, we consider the domain of science and technology where the concepts of needs and beliefs are usually quite foreign. The primary goal of this discussion is to illustrate the centrality of needs and beliefs to decisions and activities in science and technology.

Discussions in pursuit of this goal serve several purposes. First and foremost, the generality of the Needs–Beliefs–Perceptions (NBP) Model is illustrated. By arguing toward the same conclusions from point of view of culture and religion (Chapter 3) and from the perspective of science and technology (Chapter 4), we are able to illustrate the commonality of needs and beliefs. Thus, unlike the distinctions in C. P. Snow's *Two Cultures* (Snow, 1965), we conclude that the psychological phenomena underlying the "soft" disciplines and the "hard" disciplines are much the same.

This chapter also serves to help likely readers of this book relate to the phenomena of interest. Readers of this book are likely to be similar to readers of *Design for Success* (Rouse, 1991) and *Strategies for Innovation* (Rouse, 1992). These books have primarily appealed to technology-oriented practitioners, managers, and executives whose education is predominantly in science and technology.

The discussions in this chapter bring the concepts of needs and beliefs close to home for these types of readers.

A third purpose of this chapter is to provide background material for later discussions. In Chapter 8 in particular, where sociotechnical disputes are discussed, it is important to realize that these types of dispute cannot be creatively resolved if one side of the dispute (i.e., the scientists and technologists) feels that it has absolute truth on their side, while the other side has only emotions and politics. If the scientists and technologists realize that needs and beliefs also underlie their claims, it may be possible to avoid the typical tenor with which such debates are often pursued.

This chapter begins by discussing the nature of science and technology. This discussion sets the stage for exploring differences among various disciplines in science and technology. This material provides a basis for considering needs, beliefs, and perceptions in the context of science and technology. This chapter concludes by coming full circle and discussing cultures and religions in science and technology.

NATURE OF SCIENCE AND TECHNOLOGY

This section provides a broad perspective on science and technology as a human endeavor. The objectives of science include understanding of physical, behavioral, and social phenomena, and communication of this understanding. Technology is concerned with application of this understanding to practical ends.

These are admirable goals, or perhaps ideals. However, it must be recognized that science and technology pursuits are inherently human pursuits. Consequently, science and technology are subject to the same human limitations and biases that affect all other endeavors. I hasten to note that I do not view this conclusion as disappointing or negative. Having been trained as a scientist and technologist, I find the human nature of these endeavors to be fascinating, much more interesting than the stereotypical view.

History of Science

It is useful to start by considering science and technology from a historical perspective. To this end, we begin with a brief review of the history of science and technology, drawing on the review by Ravetz (1985).

For the most part, science as it exists today can be called European science. This is due to the confluence of the Renaissance, Reformation, Enlightenment, and Industrial Revolution in Europe. These eras provided the impetus for science in Europe to mature.

The roots of European science start in Greece where, as early as the 5th and 6th centuries B.C., several great mathematicians emerged. In the 4th century B.C., Plato made contributions to mathematics and, during the same time period, Aristotle contributed to biology and medicine. By the time of Jesus, however, Roman dominance with its emphasis on magic and related superstitions led to the withering of science.

Renewed activity in science emerged in the 12 century A.D., particularly in those areas where the influence of Islam was greatest. The Golden Age of Islam, noted in Chapter 3, coincided with the low point of European culture outside of those areas influenced by Islam. In the East, Chinese technology was more advanced than Europe up until the Renaissance, which started in the 14th century. By the time European missionaries arrived in China in the 16th century, China was at low ebb.

European expansion began with the Renaissance. By the late 15th century, mining, metallurgy, and trade began to grow. At roughly the same time, printing was invented and the explorations of the Spanish and Portuguese began. By the mid-16th century, the Reformation had occurred and science had rebounded.

The 17th century saw the rise of natural philosophy, with Bacon, Descartes, and Galileo as its prophets. This philosophy saw nature as devoid of human and spiritual properties. This position contrasted with the earlier scientific philosophy which emphasized finding God's laws. In the latter part of this century, Newton's *Principia* was published (1687) and scientific societies emerged.

By the end of the 18th century, the Enlightenment, with its struggle against church dogma, provided the intellectual origins of the French Revolution. Many scientists, particularly mathematicians, played central roles in this revolution. During the same period, the Industrial Revolution emerged. This led to an industrial demand for science.

The Industrial Revolution also involved dramatic changes in technology. The early 18th century saw increased mechanization, which replaced reliance on individual craftsmen. By the 19th century, interchangeability of parts and tools became common (Voelcker, 1988).

Organized research began to grow in the 19th century, particularly in Germany. Great strides occurred in physics, chemistry, and biology. This progress provided the foundations for the reductionism that has dominated the 20th century, with its emphasis on tightly controlled experiments involving phenomena that are very small in scope. The result is very crisp data, but often at a cost of not gaining understanding of more pervasive phenomena.

Currently, science is no longer the sole province of the lonely researcher in the white lab coat. Science is now interwoven with industry, defense, and politics. Recalling Neibuhr's analysis (Neibuhr, 1929), we might say that the many sects of science have now become a church, or churches, with increasing emphasis on preserving the institution, socializing its young, and fostering its own agenda.

Philosophy of Science

It is important that we consider the nature of scientific inquiry and why it is done. As is later discussed, the nature of scientific activities strongly influences needs and beliefs in science. Similarly, needs and beliefs in technology are also influenced since the education of engineers typically includes a hefty component of science.

On the surface, it appears that scientists collect much data and manipulate numerous equations. Indeed, for the novice and some journeymen scientists, their greatest joy in their work is often exercising their skills in data collection and analysis, and formulating and manipulating equations. However, as physicist John Ziman (1968) has noted, "The aim of science is understanding, not the accumulation of data and formulae."

The desired understanding involves explanations of physical, behavioral, and social phenomena. Psychologist Zenon Pylyshyn (1983) states that the goal of explanation is to produce relevant generalizations. A key word here is "relevant." Beyond the skills of data collection and equation manipulation, scientists have to decide what is relevant. Later discussion indicates the social nature of such decisions.

Biologist Robert Rosen (1986) asserts, "The main task of theoretical science is precisely to bring causal relations between events (in the external world) into a congruence with inferential relations between propositions (in a formal system) describing those events." Thus, science inherently involves constructing a representation in a "model world" and comparing its properties and behaviors to those of the real world. In this process of creating abstractions of the real world, physicist and mathematician Alfred North Whitehead (1925) cautions that we have to be careful not to mistake our abstractions for concrete reality.

John Casti (1988), an engineer and mathematician, draws upon the work of Joseph Campbell discussed in Chapter 3 to assert that science, and its models, involves myths that are predictive, empirically testable, and cumulative. Casti draws a parallel between the theories and models of science and the theologies and dogmas of religion. His goal is not to denigrate either of these points of view. Instead, he is attempting to illustrate the true nature of scientific inquiry for the purpose of determining the state of our knowledge of various phenomena.

To summarize, the goal of science is understanding and explanation. This inherently involves viewing the world through abstractions that, however useful, are not identical with physical reality. This conclusion quite naturally leads to the question of how abstractions are chosen and why one is chosen over another.

Modeling

Scientific explanations and technological creations are often characterized in terms of models. For our purposes, a model can simply be defined as an explicit representation of how something works. The "something" may, for example, be atomic particles, gene mutations, microelectronic circuits, or decision-making processes.

There are three broad classes of models (Rouse, Hammer, and Lewis, 1989). *Experiential* models are those gained by experience. Such models are useful to the extent that the past is a good model of the future. Even if this is not a good assumption, the natural tendancy of most humans is to adopt this assumption, at least implicitly, until the evidence overwhelmingly points to a need to rethink premises. As is discussed later in this chapter, this phenomena is just as prevalent in science and technology as it is in other aspects of life.

An *empirical* model involves choosing experimental conditions and the subjects of observation in a manner that allows appropriate generalizations. In other words, conditions (i.e., foods tested) and subjects (i.e., animals used) serve as models of broader populations. A common misperception of strict empiricists is to assert that *any* empirical data is inherently of greater validity than theoretical formulations. However, if conditions and subjects are not good models of the real-world phenomena of interest, then any data resulting is of little value.

An *analytical* model is usually developed by starting with explicit assumptions of "first principles" (e.g., Newton's laws) and then deriving an overall structure. This typically leads to construction of a computational representation. Various characteristics of this representation are then computed. These computed characteristics are then compared to the measurements that result when these characteristics are assessed for the real phenomena being modeled.

For any particular phenomena, there are an infinite number of models that can produce similar phenomena. How does one choose among these possibilities? John Casti (1988) argues that good models are simple, agree with most of the available data, and are structurally compatible with any knowledge of underlying phenomena. Clark Glymour and his colleagues (1987) say, "The best models are those that imply patterns or constraints judged to hold in the population, that do not imply patterns judged not to hold in the population, and that are as simple as possible."

These two sets of criteria are similar. Both the engineer and mathematician (Casti) and the psychologist and philospher (Glymour) exhibit the same set of beliefs about the nature of good models. This set of beliefs is explored in more detail later in this chapter.

There are many ways to describe how models are developed. A simple description serves our purposes (Rouse, Hammer, and Lewis, 1989). First, based on an understanding of the phenomena of interest, a representation is chosen. Typically, the choice reflects both the modeler's understanding of the phenomena *and* the modeler's experience and preferences among alternative representations. Thus, the modeler chooses the glasses through which to view the world, and these glasses heavily influence the perspective gained.

Most models have parameters or free variables within them (e.g., elasticity of the material involved or reaction times to stimuli presented). Often, an initial set of data is used to estimate these parameters. These estimates are then incorporated into the model. Next, the model is used to predict likely values of new observations. New data are collected. Predictions and observations are compared. For good models, the discrepancies are small.

This process sounds straightforward. However, bounded human rationality and combinatorial reality limit our modeling abilities (Glymour et al., 1987). Several types of modeling error can result (Rouse and Hammer, 1991). Incomplete models are missing important structural elements. Inaccurate models need better parameter estimates. Incompatible models are expressed in terms not measurable for the phenomena of interest. Incorrect models are completely off base. Other types of modeling errors include over- and underspecification of model inputs and outputs. Clearly, there are a variety of ways that a modeling effort can go awry.

These possibilities naturally lead to concerns for our abilities to eventually arrive at the "truth." This issue is pursued in depth later in this chapter. However, it is useful to provide some insight at this point.

Consider the way in which science deals with uncertainty. We have tended to view apparently unpredictable phenomena as subject to random disturbances that, by definition, cannot be predicted. Recent thinking, however, has changed this view. The field of chaos theory has, in the past 10 years or so, provided new explanations of apparent randomness (Gleick, 1987). It has been found that relatively simple nonlinear phenomena—those whose response to multiple inputs is not simply the sum of the responses to each of these inputs—can be extraordinarily sensitive to their initial conditions. Consequently, what has been characterized as random behavior may often only reflect our ignorance of initial conditions or our inability to measure initial conditions with sufficient accuracy to be able to make reasonable predictions.

The implication of this conclusion is that *our* inabilities, not the nature of the phenomena, are what limit the arrival at "truth." There are many physical, behavioral, and social phenomena for which such limits are likely to exist. Casti (1990a) discusses several of these in depth. Later in this chapter, the ways in which such limits affect needs and beliefs are considered.

Empirical vs. Axiomatic Methods

One of the central tensions in the scientific and technological community is between experimental and mathematical traditions (Kuhn, 1977). This tension involves a debate about the relative merits of empirical and axiomatic methods. While it is not usually discussed in modeling terms, this debate involves conflicting perspectives on empirical vs. analytical models.

John Ziman (1968), a physicist, argues that the basic substance of theory is reasoning, not just data. In other words, empirical data are not inherently valuable.

Their value lies in the nature of the theory or model that dictated the data be collected. Such data are valuable only in reference to this theory or model.

In contrast, psychologist Robin Hogarth (1985) says, "Science cannot depend too heavily on axiomatic reasoning since, by definition, this only applies with restricted, closed worlds that lack the open systems characteristics of our everyday reality." Thus, we have to be careful to not confuse our "model world" with the "real world." It can be argued, however, that we have no choice but to view reality through our models.

Arthur Koestler (1978) takes the debate to another plane when he asserts, "Experimental evidence can confirm certain expectations based on a theory, but it cannot confirm the theory itself." Using the glasses metaphor introduced earlier, this argument suggests that data can confirm predictions made with particular glasses on, but they cannot confirm that we are wearing the best glasses. Of course, I hasten to note, data can prove that we are definitely wearing the wrong glasses. Nevertheless, our models heavily influence what we expect, what data we collect, and how we interpret these data.

The conflict between strict empiricism and ardent axiomatic thinking is very evident if one moves back and forth across the border between behavioral and social sciences, and the physical sciences and engineering. Having spent the last 25 years making frequent border crossings, I have seen many manifestations of this conflict.

For example, I have been involved with numerous multidisciplinary studies of complex problems such as piloting aircraft, controlling power plants, and managing enterprises. The research teams in such studies typically included behavioral scientists (e.g., cognitive, experimental, industrial, or social psychologists) and engineers (e.g., aeronautical, electrical, industrial, mechanical, or systems). The natural inclination of behavioral scientists is to pursue issues experimentally. Engineers, on the other hand, are naturally inclined to think in terms of models. This is not surprising, considering the training usually received by behavioral scientists and engineers.

Conflicts arise when the results of experiments and modeling efforts are evaluated. Behavioral scientists tend to think that data are "real," while models outputs are "artificial." However, as noted earlier, experimental conditions and subjects are models in themselves. Further, statistical analyses of data are inevitably model based. Hence, behavioral scientists are modelers also. However, they do not think of themselves in this way.

Engineers have a tendency to think that behavioral scientists' tightly controlled experiments are simplistic and irrelevant, particularly if the experimental subjects are the ubiquitous college students recruited from introductory psychology courses. Engineers are inclined to build elaborate models and then, perhaps via simulation, experiment in the resulting "model world." In the process, they risk losing track of the human behaviors that they were attempting to model.

The occasional distain that behavioral scientists and engineers exhibit for each other's methods is, I think, a natural result of their respective educational experiences and the contexts within which they are used to working. Usually, they are very skilled in their own methods and reluctant to entertain the possibility that other methods may be more useful for the problem at hand. Consequently, they tend to impugn the validity, acceptability, and viability of the other methods. Later in this chapter, we consider the nature of this behavior in terms of needs and beliefs.

Sociology of Science

The striking differences among academic subcultures is elaborated in C. P. Snow's classic *Two Cultures* (Snow, 1965). In this frequently cited work, Snow contrasts the arts and humanities with the sciences. He also discusses the importance of narrowing or bridging the gulf. Several of Snow's distinctions can be applied to a finer-grained analysis of what I have called the "N Cultures" of disciplines within science and technology (Rouse, 1982). These distinctions are elaborated later in this chapter.

A variety of people have explored the social system of science. John Ziman (1968) discusses science in terms of its role in the broader social system, namely, to produce "public knowledge." Considering science as a social system itself, without doubt the book that has had the most impact is Thomas Kuhn's *The Structure of Scientific Revolutions* (Kuhn, 1962). It contrasts normal science and scientific revolutions.

Normal science is the activity that occupies almost all scientists. This activity involves elaboration of accepted paradigms. Paradigms are accepted models of classes of phenomena that are recognized by the scientific community as solving a critical set of problems better than any competing models. Most scientists within any particular community work on elaborating the attributes and nuances of the reigning paradigm in that community.

A scientific revolution occurs when a new model appears that substantially overcomes one or more of the limitations of the reigning paradigm. At first, the community is likely to dismiss the new model. However, in some cases, the new model eventually gains support and, at some point, quickly takes over as the reigning paradigm. A good example of this phenomenon is the emergence of quantum mechanics in physics.

This perspective of science easily leads to the conclusion that current scientific "truths" are, by no means, absolute—most of these truths are likely to change eventually. Reigning paradigms compete with aspiring paradigms and, occasionally, aspirants succeed. John Casti (1989) characterizes several of these competitions in his fascinating book *Paradigms Lost*.

Since normal science is the essence of mainstream science, with revolutions the exception, it should not be surprising that nonmainstream efforts encounter many barriers. For example, unusual results may be rejected out of hand, especially by anonymous reviewers, because they do not elaborate and support the accepted paradigm (Goleman, 1987). My experiences in publishing scientific articles in a wide range of journals are that reviewers have great difficulty supporting papers that employ nonmainstream research methodologies, even in situations where the mainstream methodologies are not appropriate.

It would seem that some areas would be immune to these types of subjectivity. However, I have yet to find convincing examples. For instance, DeMillo, Lipton, and Perlis (1979) clearly illustrate the ways in which social consensus affects the acceptability of mathematical proofs. This does not lessen the value of the proofs. It simply underscores the extent to which absolutes are lacking.

The subjectivity of the social system of science also interacts with the subjectivity of the broader social system. Sandra Scarr (1985) presents several examples that illustrate how social beliefs and sociocultural biases affect what researchers hypothesize and what they find. For instance, it is unacceptable to hypothesize that differences in gender, ethnic origin, or race cause differences in job performance or likely career achievements. Our society finds the pursuit of such possibilities to be divisive. The important point, from the perspective of needs and beliefs, is that science should not be viewed as truth-seeking unfettered by human proclivities to be influenced by needs and beliefs.

Psychology of Science

Beyond the social setting of science, there is the activity itself. There is a myth that the "scientific method" completely defines this activity. In fact, it is much more complicated and subtle.

Science involves the interleaving of experience, methods, communication, and critique (Koestler, 1978; Glymour, et al., 1987). Experience provides the basis for hunches and hypotheses about what is important and potential underlying phenomena. Methods provide the means for pursuing these notions. Communication involves presenting a line of reasoning and results to the scientific community. Critique comes from often anonymous reviewers of these communications.

The process of discovery in science can be viewed as puzzle or problem solving, particularly for normal science as defined earlier (Kuhn, 1962). Problem-solving models have been developed that mimic many of the human behaviors in this discovery process (Langley et al., 1986). Such models portray this process as being much more mechanical than it is, but nevertheless serve the important purpose of showing that science can be described by drawing upon the repertoire of behaviors exhibited by all humans.

Metaphors are common to much of scientific reasoning and theorizing. Physics has long had its mechanical models of phenomena ranging from cosmic to elementary particles (Cole, 1987). Biology has adopted information-processing metaphors (Sheldrake, 1986; Casti, 1990a). Psychology has tended to adopt as metaphors whatever the dominant technology is at the time (Gentner and Grudin, 1985). Within social science in general, there is a tendency for "armchair" metaphors to dominate rather ad hoc theorizing by those not trained in these fields (Ziman, 1968; Diamond, 1987).

Of particular importance to the discussions in this chapter is not the utility of any specific metaphors, but the apparent need to use them. Whether people are thinking about and discussing culture, religion, science, or technology, there is a need to communicate their ideas by referencing them to some commonly understood phenomena. As Joseph Campbell (1988) notes, however, it is important that people, scientists and otherwise, not confuse their metaphors with reality.

The psychological limitations and biases that affect all people also affect scientists and technologists. The phenomenon of *selective attention*—filtering out disagreeable concepts and results—is prevalent in science, particularly for information that does not support the reigning paradigm (Cole, 1987). Scientists exhibit the *confirmation bias*—only seeking information that supports their hypotheses—when they design experiments that do not entertain the possibility of their theories being wrong (Greenwald et al., 1986). As might be expected, *subjective interpretation* of experimental results and statistical analyses is also an inherent aspect of science (Berger and Berry, 1988).

The critical point is not that science is flawed, but that it is a human activity and, as such, it is subject to the same human foibles that attend all human activities. Science is very good at trying to minimize the impact of these problems. However, they cannot be eliminated. The greatest risk is that we think that they are eliminated and act as if we have objective, absolute truths in hand.

Summary

This section has served to show that endeavors in science and technology are not as pristine as they are often viewed. These endeavors involve real people doing highly specialized work. These people are well trained and highly motivated. Nevertheless, they are subject to the same human limitations that all people are.

The discussions in this section and these conclusions provide a basis for considering the needs, beliefs, and perceptions typically found in science and technology. By understanding relationships among needs, beliefs, and perceptions, it is likely that we can get below the surface of, for example, sociotechnical disputes. In addition, for the scientists and technologists who attempt to apply the concepts, principles, and methods elaborated in this book, the remainder of this chapter serves to convince them that the phenomena of concern are central to the ways in which they themselves approach their work.

DIFFERENCES AMONG DISCIPLINES

Insights into the needs and beliefs that underlie science and technology can be gained by considering the nature of the problems pursued within individual disciplines of science and technology. These problems, as well as any fundamental limits encountered in pursuit of these problems, provide the context within which needs, beliefs, and perceptions interact. As argued in earlier discussions, the problems and limits faced by any social group, scientific or otherwise, will strongly influence the needs, beliefs, and perceptions of group members.

Scientists and technologists face several common problems (Burke, 1987). First is the inductive nature of most knowledge. Generalizations must be inferred from a number of specific observations. Hence, the "laws of nature" cannot be proven deductively. From this perspective, the laws of thermodynamics, for instance, are really just well-corroborated conjectures.

Deductive proofs are only possible in situations where we, not nature, make the rules. Mathematics, computer science, and possibly theoretical physics are examples of disciplines that study model worlds, rather than the physical world (Rouse, 1982; Rouse, Cody, and Boff, 1991). In model worlds, we inherently know the "ground truth" about how the world works and, therefore, we can deduce the behaviors likely in such worlds.

Later in this section, disciplines are contrasted in terms of their relative emphases on real vs. model worlds, as well as other attributes. Before attempting such generalizations, however, it is useful to consider the nature of a variety of particular disciplines.

Biology

The central problems of biology include defining life, determining the limits of life, and exploring the possibilities of reversing life processes (Yarmolinsky, 1987). As biologists pursue the question of why particular forms of life exist, they are confronted with improbabilities rather than impossibilities. Many forms of life are imaginable, but most are very unlikely (Katz, 1987). Thus, studying the conditions under which life can emerge is limited to consideration of those conditions under which it did emerge.

Considering biological development, Casti (1990a) outlines three schools of thought. The vitalist school considers the organism to be imbued with a life force transcending its material composition. The organicist school characterizes the organism as a whole, rather than simply as the sum of its parts. Finally, the mechanistic school claims that life can be explained in terms of physical processes and their interactions.

Central in the debates among these points of view are the organism's ontogeny—its own course of development—and its phylogeny, or the course of development of its species. Of particular interest are the relationships between these two types of development. A phenomenon within this area that has attracted much attention is morphogenesis, or the way in which an organism evolves to the same form or shape as other members of its species.

Rupert Sheldrake (1981) has argued for the notion of morphogenetic fields, a biological analogy of gravitational fields whereby the species as a whole influences individual members of the species to assure the appropriate form results. This controversial but respected theory purports that such fields arise from previous forms of organisms and survive the death of organisms to influence the forms of individuals not yet conceived or born. Casti (1990a) classifies this theory as belonging to the organicist school.

Another level of explanation of biological development can be found at the cellular level. At this level, we encounter what Casti (1990a) terms the "central dogma of molecular biology." Based on the work of James Watson and Francis Crick, the central dogma asserts that information in the cell flows from the DNA (deoxyribonucleic acid) to the RNA (ribonucleic acid) to protein. In this way, the cellular DNA determines the proteins produced by the organism, and consequently its development, form, and behaviors.

In Casti's (1990a) review of this theory, he states, "The rationale for the whole enterprise rests upon the very specific act of molecular biological faith asserting that every (human) trait is directly traceable to the action of the genes. This is the mechanist creed in its purest form, a belief bearing directly upon the matter of predicting biological form." Sheldrake (1986) says, "The idea of a genetic program means that biology has imported the metaphor of computer programs and informational language." He further concludes, "What could be more anthropomorphic in human modeling than to say that everything's a machine? Machines are entirely and specifically human creations."

Note how Casti discusses these competing theories using words such as faith, belief, creed, and dogma. Sheldrake claims that molecular biologists are adopting metaphors and building theories in the image of themselves, perhaps without realizing it. These observations do not imply that biological science is being other than scientific. They simply provide a glimpse at the underlying psychology of model or theory building.

Chemistry

Gensler (1987) portrays chemistry as an inductive science, laced with insights and leaps of intuition. Most principles of chemistry are summaries of the results of repeated experiences. Thus, for example, the laws of thermodynamics and conservation principles cannot be proven in the conventional sense of proof. Scientists believe these laws and principles because they are reasonable and no one has observed violations of these laws.

Ilya Prigogine (1980) has attempted to bring a deeper understanding to the laws of thermodynamics. Based on elements of chaos theory, which was discussed earlier in this chapter, he argues that the second law of thermodynamics—which predicts ever-increasing disorder (entropy) and decreasing energy—does not necessarily hold for systems that have the capacity for self-organization and creativity. He argues for rethinking our traditional mechanistic view of the world in general, and the ways in which time is considered in particular. He suggests that our static and mechanistic views in science are probably the results of our conceptualization of God in terms of an eternal, unchanging wisdom (Prigogine, 1986).

Chemistry has seen many theories rise and fall (Gensler, 1987). The phlogiston theory of burning, the vital force explanation of life, the separateness of the laws of conservation of mass and energy, and the immutability of atoms are all "truths" that have been accepted, endorsed, and eventually discarded. At the time of their acceptance, mainstream scientists believed in the fundamental nature of these theories. They were the best explanations available at the time.

Due to its inductive nature, chemistry is inevitably faced with the need to believe in truths that will undoubtably change as methods improve and enable better theories. This is not necessarily a problem, if scientists realize the inevitability of this evolution. This realization should enable scientists to keep whatever is the prevailing paradigm from becoming dogma, in the stifling sense of the term. Nevertheless, we have to recognize the compelling human need to "nail down" some basic premises rather than to continually question everything.

Physics

Physics is both an inductive and deductive science, sometimes playing by nature's rules and sometimes making its own (Bohm, 1986; Bohm and Peat, 1987; Park, 1987). In the former case (i.e., experimental physics), physical laws are induced and physical impossibilities are of interest. In the latter case (i.e., theoretical physics), mathematical laws are deduced and logical impossibilities are of concern.

Physics is a mature science in that a variety of limitations and dilemmas are understood and formalized. An example is the uncertainty principle in particle physics whereby one inherently cannot measure both position and momentum with unlimited accuracy (Heisenberg, 1958). This is due, in part, to the constructs of position and momentum not being meaningful, in the same way, at the subatomic level as they are at the macroscopic level.

Another explanation is that position cannot be measured precisely without affecting momentum, and vice versa. Another illustration of the dilemmas faced by physics concerns the nature of light. In some cases, light behaves as a wave; in others, it behaves as a particle. Physicists accept that both theories or models are "true," and the truth you employ depends on the situation of interest. Thus, the dilemma is simply accepted.

Despite such limitations and dilemmas, physicists have not been reluctant to apply their theories quite broadly. A variety of physicists have proposed Grand Unified Theories, or "theories of everything" (Cole, 1987; Davies, 1992). Theories range from the origin of the cosmos (Hawking, 1988) to quantum mechanical explanations of consciousness (Penrose, 1989). In the process of developing such theories, physicists do not hesitate to invoke concepts such as the "psychology of elementary particles" (Zukav, 1979).

More than chemistry, and perhaps biology, physics tends to be theory driven. Physics has also experienced the rise and fall of many theories over hundreds of years, or longer if Greek heritage is included. While most physicists work within the reigning paradigms, with great regularity grand explanations emerge, are popularized, and usually recede into the archives. This penchant to attempt to explain everything is common to all science, but most developed in physics.

Psychology

Psychological research is especially difficult. Psychologists must induce the nature of a complex, intentional entity. One of the difficulties is that psychologists themselves are instances of this class of entity. In the early years of psychological research, this difficulty was overlooked. In fact, introspection—asking yourself about yourself—was a common method of gathering "data."

Early in this century, James Watson (1914) introduced a new paradigm in psychology—behaviorism. B. F. Skinner (1938) embraced this paradigm and, until his recent death, was one of our most well-known psychologists.

Behaviorism, as the name implies, is only concerned with observable behavior and its relationship to observable stimuli. It does not allow, or admit any need, for any constructs associated with the mind, mentality, and so on. Behaviorism is, in effect, pure empiricism. Behaviorism rescued psychology from the introspectionists. Information-processing theorists, such as Herbert Simon (1991), rescued psychology from the behaviorists. Using computers, both as metaphors and tools, information-processing models were developed as early as the 1950s to account for the mental aspects of human behavior that the behaviorists had consciously ignored.

As physics did earlier in this century, psychology has encountered a variety of limitations and dilemmas. For example, when people learn a new concept, do they learn it gradually and suddenly exhibit the new knowledge, or do they learn it at the moment that they display the new knowledge? It has been found that it is impossible to determine whether learning is gradual or "all or none" (Rumelhart, 1967).

As another illustration, is human information processing serial or parallel?

Townsend (1974) shows that this question cannot be answered. Are mental images stored as images or constructed from stored propositions? Anderson's (1978) assertion that this question cannot be answered led to a lengthy debate (Anderson, 1979; Hayes-Roth, 1979; Pylyshyn, 1979; Johnson-Laird, 1983). How do people solve problems, such as puzzles or games? Simon (1975) and Einhorn, Kleinmuntz, and Kleinmuntz (1979) show that a variety of models can produce the same problem-solving behavior and, therefore, that it is impossible to infer what specific model is being employed.

Other dilemmas in psychology include the possibility that probing people's memory—by asking them to remember something—can affect what they remember (Loftus and Palmer, 1974; Faust and Ziskin, 1988). Therefore, many issues cannot be resolved conclusively by simply asking people. Another limitation is the impossibility of consistently rational social choice (Arrow, 1963). There is no procedure that can assure that a group of consistent and rational individuals can always produce a group decision that is also consistent and rational.

As physics attempts to produce Grand Unified Theories, psychology attempts to produce comprehensive models of the mind. Recent entrants in this competition are Roger Penrose's (1989) model based on quantum mechanics and Daniel Dennett's (1991) model drawn from psychology and philosophy. Herbert Simon's recent autobiography (1991) chronicles the development of his work, and that of his colleagues, in this direction. A variety of the competing models are reviewed by Rouse and Morris (1986).

Psychology is a much younger discipline than physics. However, it has quickly encountered similar problems and engendered similar aspirations. The limitations outlined in this section are psychology's equivalent to physics' uncertainty principle and wave-particle dilemma. Psychology's push for models of the mind reflect the same types of aspirations motivating Grand Unified Theories. It seems that similar needs and beliefs underlie these different efforts. This possibility is explored later in this chapter.

Mathematics

The discipline of mathematics operates totally within a world that it has created—a model world. Consequently, deductive proofs are possible. In fact, proofs, as opposed to experimental data, are the currency of mathematics. Nevertheless, mathematicians face a variety of daunting problems (Guillen,1983; Davis, 1987).

For dynamic phenomena, mathematicians must cope with sampling limits, apparent discontinuities, and apparent randomness. Sampling limits dictate that signals be measured with a sufficient frequency to enable estimating signal values between samples accurately. Discontinuities have been considered in terms of catastrophe theory which depicts situations where very small changes of parameters result in dramatically different responses (Mathjay, 1985). Apparent randomness has been studied using chaos theory, and it has been shown that response patterns of nonlinear systems that had been judged to be random could be attributed to small variations of initial conditions (Gleick, 1987).

Mathematicians also must deal with problems of computational complexity (Lewis and Papadimitriou, 1978). It has been shown that for one class of problems the time to compute a solution increases exponentially with the size of the problem—in other words, computation time becomes infinite much sooner than the size of the problem becomes infinite. It has also been determined that a variety of problems are not solvable (Godel, 1962; Davis, 1965). In some cases, it has been proven that a problem can be solved, but no one has been able to find the solution.

Mathematicians set the rules of the game, decide what problems are important, and then proceed to prove theorems about the nature of problems and solutions, as well as what can and cannot be solved. This does not imply that mathematics is at all arbitrary. Setting the rules of the game and defining the most important problems started with the Greeks over 2,000 years ago. As a consequence, traditions are very strong regarding the nature of discourse in mathematics and what constitutes a contribution.

Summary

This section has provided a very rapid tour of a variety of scientific disciplines. The goal was to provide a flavor of some of the differences among disciplines. Such differences have been explored by numerous people. Snow's (1965) classic on "two cultures" was discussed earlier. Davis and Park's (1987) compendium includes many insightful papers on the limits faced by different disciplines.

It has long seemed to me that there are systematic differences among disciplines, both in terms of the nature of the problems pursued and methods employed. This hypothesis led to a paper subtitled "N Cultures" (Rouse, 1982). More recently, the scheme presented in this paper was expanded (Rouse, Cody, and Boff, 1991). This scheme can be used to summarize much of the discussion in this section.

Figure 4.1 illustrates relationships among a variety of disciplines. One dimension is the real- vs. model-world distinction that was noted in earlier discussions. The second dimension concerns internal vs. external phenomena. Internal phenomena are those in which the observer is also a participant, which inevitably flavors the ways in which such phenomena are viewed. External phenomena are those in which either the observer is not a participant or those in which participation is easily externalized (e.g., physiology).

Summarizing our analysis of the meaning of positions in the space defined by Figure 4.1—see Rouse, Cody, and Boff (1991) for detailed discussion—disciplines toward the right and toward the top tend to emphasize decomposition methods, while those toward the left and toward the bottom emphasize holistic

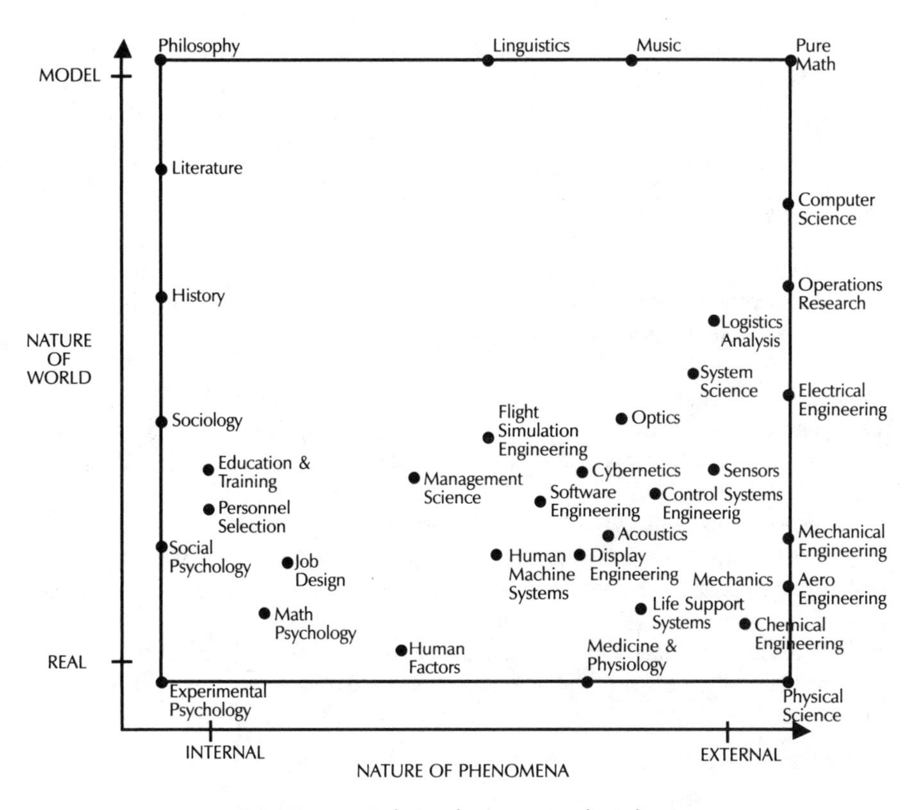

FIGURE 4.1. Relationships among disciplines.

methods. The reason is quite simple: external phenomena in the model world can often be taken apart and inspected. In contrast, internal phenomena in the real world usually cannot be decomposed in a way that retains the essence of the parts.

Another dimension of this scheme for comparing disciplines is problem scope. For small problems, both holistic and decomposition points of view favor analytical methods. For moderately sized problems, holistic approaches dictate empirical methods, while decomposition approaches usually can still retain analytical methods. For large problems, even decomposition approaches require empirical methods. Succinctly, deduction becomes untenable much more quickly for problems that have to be approached holistically.

This analysis has important implications for how different disciplines view problems and their natural tendencies for how problems should be solved. I have found that scientists' views of all of nature are highly influenced by their views of the small piece of nature about which they are expert. Thus, for example, those steeped in decomposition tend to apply reductionism to all problems. Analytical modelers try to describe everything in terms of equations and computer programs. Holistic empiricists see experimental data as the saving grace of any effort.

NEEDS, BELIEFS, AND PERCEPTIONS

The context of scientific and technical disciplines strongly influences the needs and beliefs of scientists and engineers. This context includes the problems addressed, methods employed, limitations encountered, and disciplines' relationships to broader society. These contextual aspects of science and technology influence needs and beliefs just as the context of culture and religion influences needs and beliefs. Consequently, as Ziman (1968) notes, within science, "there is far more faith, and reliance upon personal experience and intellectual authority, than the official doctrine will allow."

Of course, this really should not be surprising. The psychological underpinnings of perceptions and decisions, in terms of needs and beliefs, are similar for all people, regardless of whether one is an artist or a rocket scientist. This realization is particularly important for addressing the four archetypical innovation problems that are discussed in later chapters. If scientists and technologists understand how their needs and beliefs affect their perceptions and decisions, they will be in a much better position to address these types of problems creatively.

Needs

Figure 4.2 summarizes the needs identified in the discussions of science and technology in this chapter. As with the set of needs identified in Chapter 3—summarized in Figure 3.5—this list is roughly ordered according to Maslow's (1954) hierarchy. Physiological and safety needs are not reflected in Figure 4.2. While scientists and technologists do need to eat and feel safe, meeting of such needs is not usually viewed as central to the essence of science and technology.

The first set of needs in Figure 4.2 reflects the need to belong to a scientific and/or technological community. These needs not only relate to the general human need to belong, but also are central to succeeding in the professions associated with science and technology. Thus, budding scientists and technologists are well served by pursuing mainstream R&D topics, for example, and assuring that their efforts are communicated to their disciplinary community and well received by this community.

The next set of needs in Figure 4.2 relates to the use of hard-won skills. Usually, scientists and engineers spend years gaining very specialized skills such as mathematical modeling, numerical analysis, and experimental design and analysis. They need to use these skills and feel that, as a result, they accomplish something

- Need to belong to discipline
- Need to stay in paradigm
- Need to play by the rules of the game
- Need to communicate
- Need for social consensus
- Need to comply with broader society

- Need to employ particular methods
- Need to exercise skills
- Need for skills to be valued
- Need to solve puzzles/problems
- Need to have an impact

- Need for structure
- Need to represent
- Need to predict
- Need for control
- Need for autonomy
- Need to be right

- Need for natural causes
- Need for some immutable premises
- Need for continuity
- Need for metaphors

- Need to understand
- Need to explain
- Need to explain everything
- Need to accept dilemmas

FIGURE 4.2. Summary of needs in science and technology.

that is valued. Consequently, someone may reject a problem as unimportant because solving the problem does not require his or her specialized skills.

I have experienced this frequently. Until I realized why it was happening, I was involved in what seemed like endless debates about the importance of particular problems. It is often more productive to spend this time showing the reluctant individual how his or her skills are important to solving the problem at hand.

The third set of needs in Figure 4.2 concerns several natural tendencies when scientists and technologists approach problems. A search for structure and representation of this structure often dominate the problem solving. Based on the resulting representation, there are subsequent needs to predict the evolution of the phenomena of interest and to be able to control this phenomena. There is usually a strong push to prove that the current hypotheses about the problem and phenomena are right. This set of needs is common to virtually all disciplines within science and technology.

The fourth set of needs in Figure 4.2 reflects the types of solutions expected. Natural causes of problems and phenomena are assumed. Standard sets of assumptions or premises are usually invoked—typically drawn from the reigning paradigm—because it is untenable to leave all issues open. These premises are then usually treated as if they were immutable, even though everyone realizes that they are not. There is also a need for continuity in the sense that one can safely build on past results, especially those that have been in the mainstream of the reigning paradigm.

There is, in all disciplines, a strong need to employ metaphors. As noted earlier, technological metaphors are often adopted. As a result many theories, at least on the surface, tend to look like the technology that was predominant when the theories emerged. Once a paradigm matures, there is a risk that many of the "normal science" toilers will forget that the models they are elaborating are only metaphors.

The fifth and final set of needs in Figure 4.2 reflects the needs that drive many scientific quests. As noted in the beginning of this chapter, needs to understand and explain are the raison d'être of science. We also noted in the discussions of different disciplines the penchant to try to explain everything with, for example, Grand Unified Theories and theories of the mind.

Finally, mature scientists and technologists recognize the need to accept dilemmas. No paradigm or set of paradigms provides credible explanations for everything. Sometimes, as with the nature of light, more than one explanation is needed. These multiple explanations are not necessarily compatible. However, accepting such dilemmas enables progress. Further, such acceptance is part of the process of understanding the limits of science.

As noted earlier, the needs listed in Figure 4.2 can be viewed as ordered in a manner similar to Maslow's (1954) hierarchy. These needs can also be considered in terms of McClelland's (1987) motives. Clearly, the achievement motive is reflected in many of these needs. Some of the affiliation motive can be associated with the need to belong to a disciplinary community. Also, a bit of the power motive might be seen in the needs to predict and control.

Beliefs

Figure 4.3 provides a summary of the beliefs identified in the discussions in this chapter. The first set of beliefs is associated with the methods and tools of science and technology. While many cultural and religious beliefs are related to conclusions about the world, beliefs in science and technology relate more to the means

- Belief in scientific methods
- Belief in abstractions
- Belief in modeling
- Belief in experimentation
- Belief in proofs

- Belief in underlying order
- Belief in natural philosophy
- Belief in mechanistic world
- Belief in reductionism
- Belief in simplicity

- Belief in the paradigm
- Belief in model world

- Belief in importance of understanding
- Belief in attainability of truth
- Belief in multiple truths

- Belief in one's reasoning abilities
- Belief in generality of one's expertise

FIGURE 4.3. Summary of beliefs in science and technology.

to reach conclusions. From this perspective, scientists and engineers believe conclusions because of their beliefs in the means whereby the conclusions were reached. The second set of beliefs is concerned with the nature of the world. Scientists believe that there is an underlying order and that they can use their methods to uncover it. Further, they believe that this order has a physical, as opposed to mystical, basis. Machine metaphors dominate beliefs about the nature of the world. This enables the belief that the machine can be taken apart by reductionism, and that the whole can be explained in terms of explanations of the parts. Moreover, there is widespread belief that the simplest, acceptable explanation is most likely the best.

As shown by the third set of beliefs, there is a strong tendency to believe in the reigning paradigm, to accept it as truth, at least from an operational point of view. Associated with this belief, there is often a tendency to view the model world as synonymous with the real world. Consequently, as Campbell (1988) has noted for religions, metaphors can become reality and distinctions can arise that are due to conflicts of metaphors, not conflicts in reality.

The fourth set of beliefs provides a basis for valuing science and technology.

Believing in the importance of understanding and the attainability of truth enables people and organizations to invest themselves in science and technology. Beliefs in multiple truths are, I think, associated with acceptance of dilemmas.

The fifth and final set of beliefs has a pervasive impact. Scientists and engineers typically have strong beliefs in their own reasoning abilities, as well as the power of reasoning in general. They also tend to believe in the generality of their expertise, which can lead to authoritative pronouncements in areas far afield of particular individuals' expertise. This can easily lead to cross-disciplinary conflicts and disputes, as the case in the next section illustrates.

Perceptions

Relationships among needs, beliefs, and perceptions in science and technology are best illustrated in the context of examples. This section uses the NBP Template to frame the analysis of a dispute involving nuclear power. This example is indicative of the types of problems considered in Chapter 8.

The dispute of interest involved a committee of university faculty that was convened to consider how to deal with the *perceived* risks of nuclear power. The committee membership included nuclear engineers, a systems engineer (me!), psychologists, and a political scientist. The goal of the group was to produce a proposal for understanding and hopefully improving the public's perceptions of the risks associated with nuclear power plants.

The nuclear engineers on the committee quickly brought forth reams of numbers gleaned from risk analyses, fault-tree analyses, and so on. They asserted that nuclear power was much safer than many other things in life. We only had to educate the public and the problem would disappear.

The psychologists felt that the problem was, by no means, solved so easily. Using expectation and attribution theories, which we discussed in Chapter 2, they explained how people come to perceive risks. They suggested that the models and analyses of the nuclear engineers would have little impact on the general public's perceptions.

The political scientist focused on how public decisions get made. She discussed regulatory processes, the roles of hearings, and so forth. The ways in which the public's perceptions of risks influence these processes were also noted.

During our lengthy first meeting, the discussion quickly led to our having a dispute of our own! There were clearly two "camps" on the committee—the pronuclear camp and the doubters. As I recall, this dispute resulted in the committee never producing the proposal for which it was convened. In the remainder of this section, the NBP Template is used to analyze this dispute.

Figure 4.4 is the NBP Template for this analysis. The pro-nuclear position is summarized in terms of the positive perceptions of viability, acceptability, and validity. People in this position believed in their models and analyses, believed

ATTRIBUTE	PERCEPTIONS	BELIEFS	NEEDS
Viability	Positive	Economic and environmental benefits of not burning fossil fuels far exceed costs	Need to have skills valued
	Negative	Psychological costs and remote but immense costs of consequences far outweight benefits	Lack of inherent need for technology to be valued
Acceptability	Postive	Risks are acceptable if calculated probabilities show risks to be smaller than other acceptable risks	Need to predict and control, as well as be right
	Negative	Even though improbable, possible consequences are unknown, far-reaching, and long-lasting	Need to have skills valued
Validity	Positive	Predictions of models and results of analyses show that risks are very small	Need to predict and control, as well as be right
	Negative	Lack of rigorously collected operational data undermines credibility of models and analyses	Need to have skills valued

FIGURE 4.4. Example NBP Template.

that the calculated risks were clearly acceptable, and believed that the benefits of nuclear power far exceeded its costs. At a deeper level, these people needed to feel that they could predict and control nuclear-related phenomena and that their models were right. They also needed to feel that nuclear engineering expertise was valuable because nuclear power plants were valuable.

In contrast, the doubters' position is represented by the negative perceptions of viability, acceptability, and validity. People in this position believed that the intangible psychological costs of perceived risks, as well as potential immense costs of low-probability consequences, far outweighed potential benefits. This is, in part, related to their lack of inherent need for the nuclear power technology to be valued.

The doubters also believed that important psychological and social phenomena had been ignored and that too much reliance had been placed on purely analytical methods. The modeling efforts and risk analyses had occurred without the benefits of their types of expertise. The doubters needed to have their skills valued in approaching problems associated with phenomena about which they felt they were expert.

Hence, the dispute was not just about the facts of nuclear power. Peoples' overall beliefs were heavily influenced by their specific beliefs about those phenomena for which they had some particular interest and expertise. Furthermore, they had inherent needs to have their expertise valued and be consulted in those areas where they felt their opinions mattered.

Based on the analysis summarized in Figure 4.4, how might this dispute have been resolved? How might this group have moved beyond the dispute and become productive? I think that the first step would be to get below the surface perceptions. While the two camps differed about the value of nuclear power, they probably could have agreed on how to go about making sure that the problem was formulated appropriately. An important aspect of this formulation would have been to agree that the prime concern was perceptions about nuclear power rather than the facts about nuclear power. This would have required the pro-nuclear camp to agree that perceptions matter, regardless of whether they are correct. Similarly, the doubters would have had to agree that the facts of nuclear power also matter. The question then would have become one of determining how perceptions might have been appropriately changed so as to enable progress in this dispute.

As I noted earlier, this committee did not get below the surface. The analysis just presented is based solely on my memories of the experience. Further, not being able to probe needs and beliefs that people had several years ago, it is quite possible that some of my inferences are wrong. Nevertheless, this analysis does illustrate how needs and beliefs in science and technology can influence perceptions. It also shows that what might be viewed as purely disputes about facts can, below the surface, involve other issues and concerns.

Summary

This section has emphasized the needs, beliefs, and perceptions associated with a wide range of scientific disciplines. To the extent that these disciplines underlie technology-oriented disciplines such as engineering, there are likely to be common needs, beliefs, and perceptions. However, it is important to note that science and engineering are not synonymous. Engineers, for example, also have beliefs about the importance of high quality products and needs to design such products. To a great extent, engineers have one foot in science and technology and the other foot in business and commerce. Therefore, important elements of engineers' needs, beliefs, and perceptions relate to markets, customers, organizations, and so on. These elements are considered in Chapters 6–7.

CULTURES AND RELIGIONS IN SCIENCE AND TECHNOLOGY

The last section of this chapter on science and technology concludes by bringing the discussion full circle, back to culture and religion. This chapter began by emphasizing the idea that science and technolgy can provide another view of needs and beliefs, a view rather different than that provided by culture and religion. In this section, the emphasis is on the similarities of these two perspectives.

Science as Religion

One cannot study these two points of view very long without beginning to see many similarities. As physicist Freeman Dyson (1988) comments, "Science and religion are two human enterprises sharing many common features." Another physicist, Paul Davies (1992), notes, "If a religion is defined to be a system of thought which requires belief in unprovable truths, then mathematics is the only religion that can prove it is a religion!"

Engineer and mathematician John Casti (1990b) provides a more elaborate view of mathematics:

> *I think that a lot of bright people who might in an earlier era have joined the church, have instead moved into the academic life as a way to scratch their essentially religious itch. This match-up is especially clear in mathematics, where you have things like a secret language comprehensible only to the initiated, a professed disdain for and detachment from worldly affairs, a prolonged period of training for new recruits under what amount to monkish material circumstances, "sacred missions" (i.e., famous unsolved problems) to which some members of the faith devote their entire lives, and so on and so forth. These parallels can be carried quite a bit further, accounting for a lot of what otherwise seems like completely incomprehensible behavior on the part of many of our academic colleagues.*

Psychologist David Wulff (1991, p. 36) provides similar insights into psychology:

> *Within the various psychologies some of the basic elements of religious traditions are discernible. We find, for example, sacred texts written by charismatic leaders or prophets who help to formulate the dogmas and creeds by which orthodoxy is defined. There are special rites of acceptance and initiation, and the oft-repeated rituals that are called methods and techniques. Frequently there are objects of veneration—experimental apparatus, psychological tests, electronic computers—and even objects of sacrifice, in the form of laboratory animals. Much like religious devotees, psychologists have sacred places for their various rites, including the conference halls where they gather periodically to recite their creeds and to testify, as well as the offices and classrooms where they work to win converts. More generally and profoundly, beneath these obvious forms lies a faith that adherence to the teachings of the tradition will in time bring salvation, whether it be in the form of a personal career, the health of a patient, or the transformation of society.*

Based on these characterizations, it is easy to see how the elements of culture and religion discussed in Chapter 2 can be found within science and technology. Thus, the professional societies within which many of us frequently participate are, in fact, societies in the broader sense of the term. This provides credence to the notion that the psychological underpinnings of science and religion are quite similar. While scientists and technologists may be better educated and more sophisticated than many adherents of the world's religions, a sufficiently deep level of analysis leads to the conclusion that many of the needs and beliefs that influence perceptions are common to all people.

Science and Religion

The relationship between science and religion has long been of great interest. As noted earlier in this chapter, science was initially viewed as the search for God's laws. With the Enlightenment, scientists came to emphasize the inherent differences between science and religion. In recent years, particularly among the more mature sciences, there appears to be a great interest in the possible unity of science and religion.

This interest is exemplified by numerous scientists' conceptualizations of ultimate reality. Robert Wright (1988) interviewed three well-known and respected scientists to determine their conceptualizations. Edward Fredkin, a computer scientist, characterized the universe as an immense computational machine. Edward Wilson, a biologist, expressed a sense of the ultimate as a superorganism. Kenneth Boulding, an economist, emphasized a sense of unity and total harmony.

In pursuit of the same question, physicist Paul Davies (1992, p. 214) says:

> *My own inclination is to suppose that qualities such as ingenuity, economy, beauty, and so on have a genuine transcendent reality—they are not merely the product of*

human experience—and that these qualities are reflected in the structure of the natural world. Whether such qualities can themselves bring the universe into existence I don't know. If they could, one could conceive of God as merely a mythical personification of such creative qualities, rather than as an independent agent.

Renee Weber (1986) conducted a series of interviews addressing the question of the possibility of unity between science and mysticism. Scientists interviewed included physicists David Bohm and Stephen Hawking, chemist Ilya Prigogine, and biologist Rupert Sheldrake. Mystics or sages interviewed included Buddhist scholar Lama Anagarika Govinda, Catholic priest Father Bede Griffiths, philosopher Krishnamurti, and the exiled leader of Tibetan Buddhists, the Dalai Lama. Several of the interviews were conducted with two of these individuals at once.

What is compelling about these interviews is not that everyone had the exact same conceptualization of ultimate reality—they did not. Instead, the commonality is the questions being asked and the breadth of the types of answers explored. All of these people are seeking to understand and appreciate the universal and the ways in which everything fits together, rather than how everything can be decomposed into tidy, researchable parts.

Davies (1992) discusses this trend toward unity by noting how Western theology influenced the flourishing of science in search of God's laws. In contrast, Eastern theology did not posit a divine, rational being who could legislate natural laws. While the reductionism of Western theology enabled much scientific progress, Davies suggests that the holism of Eastern theology is more compatible with modern nonlinear physics. From this perspective, it is not surprising that many well-known scientists are increasingly finding much in common with mystics who are apparently exploring similar questions of unity.

As Joseph Campbell (1988) has noted, "Science is breaking through now into the mystery dimension." Scott Peck (1978, pp. 227–228) has observed this trend and asked, "Is it possible that the path of spiritual growth that proceeds from religious superstition to scientific skepticism may ultimately lead us to a genuine religious reality? This beginning possibility of unification of religion and science is the most significant and exciting happening in our intellectual life today."

While these apparent trends are fascinating, we should be careful not to inflate expectations. As Paul Davies (1992, p. 231) reminds us, "We are barred from ultimate knowledge, from ultimate explanation, by the very rules of reasoning that prompt us to seek such an explanation in the first place." This is due to the fact that "no rational system can be proved both consistent and complete. There will always remain some openness, some element of mystery, something unexplained" (p. 167). Of course, such limitations are yet another instance of commonality between science and religion.

SUMMARY

This chapter has reviewed the history, philosophy, and nature of science and technology. The social, cultural, and religious aspects of science and technology have also been discussed. The goal of this discussion was not to equate science and technology with culture and religion. While some fascinating thoughts on this possibility have recently emerged, this book is not premised on accepting or rejecting this possibility.

Instead, the goal has been to provide substantive evidence for the hypothesis that these two domains of human endeavor have common psychological underpinnings. More specifically, the objective has been to provide support for the generality of the Needs–Beliefs–Perceptions Model, and the utility of the Needs–Beliefs–Perceptions Template. The intense discussions and analyses of Chapters 2–4 have provided a wealth of insights into needs, beliefs, and perceptions. We are now in a position to elaborate the NBP Model.

REFERENCES

Anderson, J. R. (1978). Arguments concerning representations for mental imagery. *Psychological Review, 85*, 249–277.

Anderson, J. R. (1979). Further arguments concerning representations for mental imagery: A response to Hayes-Roth and Pylyshyn. *Psychological Review, 86*, 395–406.

Arrow, K. J. (1963). *Social choice and individual values*. New York: Wiley.

Berger, J. O., and Berry, D. A. (1988). Statistical analysis and the illusion of objectivity. *American Scientist, 76*, 159–165.

Bohm, D. (1986). The implicate and the super-implicate order. In R. Weber (Ed.), *Dialogues with scientists and sages: The search for unity* (Chapter 2). New York: Routledge and Kegan Paul.

Bohm, D., and Peat, F. D. (1987). *Science, order, and creativity*. New York: Bantam.

Burke, T. E. (1987). What can be known. In P. J. Davis and D. Park (Eds.) *No way: The nature of the impossible* (pp. 294–315). New York: Freeman.

Campbell, J. (1988). *The power of myth*. New York: Doubleday.

Casti, J. L. (1988). *Alternate realities: Mathematical models of nature and man*. New York: Wiley.

Casti, J. L. (1989). *Paradigms lost: Images of man in the mirror of science*. New York: Morrow.

Casti, J. L. (1990a). *Searching for certainty: What scientists can know about the future*. New York: Morrow.

Casti, J. L. (1990b). *Personal communication*. September 13, 1990.

Cole, K. C. (1987). A theory of everything. *The New York Times Magazine*, October 18, 20–28.

Davies, P. (1992). *The mind of God: The scientific basis for a rational world.* New York: Simon and Schuster.

Davis, M. (Ed.) (1965). *The undecidable: Basic papers on undecidable propositions, unsolvable problems, and computable functions.* Hewlett, NY: Raven Press.

Davis, P. J. (1987). When mathematics says no. In P. J. Davis and D. Park (Eds.), *No way: The nature of the impossible* (pp. 161–177). New York: Freeman.

Davis, P. J., and Park, D. (Eds.) (1987). *No way: The nature of the impossible.* New York: Freeman.

DeMillo, R. A., Lipton, R. J., and Perlis, A. J. (1979). Social processes and proofs of theorems and programs. *Communications of the ACM, 22,* 271–280.

Dennett, D. C. (1991). *Consciousness explained.* Boston: Little, Brown.

Diamond, J. (1987). Soft sciences are often harder than hard sciences. *Discover*, (August), 34–39.

Dyson, F. (1988). *Infinite in all directions.* New York: Harper and Row.

Einhorn, H. J., Kleinmuntz, D. N., and Kleinmuntz, B. (1979). Linear regression and process–tracing models of judgment. *Psychological Review, 86*, 465–485.

Faust, D., and Ziskin, J. (1988). The expert witness in psychology and psychiatry. *Science, 241*, 31–35.

Gensler, W. J. (1987). Impossibilities in chemistry: Their rise, nature, and some great falls. In P. J. Davis and D. Park (Eds.), *No way: The nature of the impossible* (pp. 73–89). New York: Freeman.

Gentner, D., and Grudin, J. (1985). The evolution of mental metaphors in psychology: A 90-year retrospective. *American Psychologist, 40*, 181–192.

Gleick, G. (1987). *Chaos: Making a new science.* New York: Viking.

Glymour, C., Scheines, R., Spirtes, P., and Kelly, K. (1987). *Discovering causal structure: Artificial intelligence, philosophy of science, and statistical modeling.* Orlando, FL: Academic.

Godel, K. (1962). *On formally undecidable propositions.* New York: Basic Books.

Goleman, D. (1987), Failing to recognize bias in science. *Technology Review, 90*, (November/December), 26–27.

Greenwald, A. G., Pratkanis, A. R., Leippe, M. R., and Baumgardner, M. H. (1986). Under what conditions does theory obstruct research progress? *Psychological Review, 93*, 216–229.

Guillen, M. (1983). *Bridges to infinity: The human side of mathematics.* Los Angeles: Tarcher.

Hawking, S. W. (1988). *A brief history of time: From the big bang to black holes.* New York: Bantam.

Hayes-Roth, F. (1979). Distinguishing theories of representation: A critique of Anderson's arguments concerning mental imagery. *Psychological Review, 86*, 376–382.

Heisenberg, W. (1958). *Physics and philosophy.* New York: Harper and Row.

Hogarth, R. M. (1985). *Why bother with experiments?* Chicago: University of Chicago, Graduate School of Business.

Johnson-Laird, P. N. (1983). *Mental models: Towards a cognitive science of language, inference, and consciousness.* Cambridge, MA: Harvard University Press.

Katz, M. J. (1987). Are there biological impossibilities? In P. J. Davis and D. Park (Eds.), *No way: The nature of the impossible* (pp. 11–27). New York: Freeman.

Koestler, A. (1978). *Janus: A summing up.* New York: Random House.

Kuhn, T. S. (1962). *The structure of scientific revolutions.* Chicago: University of Chicago Press.

Kuhn, T. S. (1977). *The essential tension: Selected studies in scientific tradition and change.* Chicago: University of Chicago Press.

Langley, P., Simon, H., Bradshaw, G., and Zytkow, J. (1986). *Scientific discovery: An account of the creative process.* Cambridge, MA: MIT Press.

Lewis, H. R., and Papadimitriou, C. H. (1978). The efficiency of algorithms. *Scientific American, 238*, 96–109.

Loftus, E. F., and Palmer, J. C. (1974). Reconstruction of automobile destruction: An example of the interaction between language and memory. *Journal of Verbal Learning and Verbal Behavior, 13*, 585–589.

Maslow, A. H. (1954). *Motivation and personality.* New York: Harper.

Mathjay, A. (1985). *Foundations of catastrophe theory.* Boston: Pitman.

McClelland, D. C. (1987). *Human motivation.* Cambridge, UK: Cambridge University Press.

Neibuhr, H. R. (1929). *The social sources of denominationalism.* New York: Henry Holt.

Park, D. (1987). When nature says no. In P. J. Davis and D. Park (Eds.), *No way: The nature of the impossible* (pp. 139–159). New York: Freeman.

Peck, M. S. (1978). *The road less traveled: A new psychology of love, traditional values and spiritual growth.* New York: Simon and Schuster.

Penrose, R. (1989). *The emperor's new mind: Concerning computers, minds, and the laws of physics.* New York: Oxford University Press.

Prigogine, I. (1980). *From being to becoming: Time and complexity in the physical sciences.* San Francisco: Freeman.

Prigogine, I. (1986). The reenchantment of nature. In R. Weber (Ed.), *Dialogues with scientists and sages: The search for unity* (Chapter 10). New York: Routledge and Kegan Paul.

Pylyshyn, Z. W. (1979). Validating computational models: A critique of Anderson's indeterminacy of representation claim. *Psychological Review, 86*, 383–394.

Pylyshyn, Z. W. (1983). *Computation and cognition.* Cambridge, MA: MIT Press.

Ravetz, J. R. (1985). The history of science. *Encyclopedia Britannica, 27*, 30–39.

Rosen, R. (1986). Causal structures in brains and machines. *International Journal of General Systems, 12*, 107–126.

Rouse, W. B. (1982). On models and modelers: N cultures. *IEEE Transactions on Systems, Man and Cybernetics, 12*, 605–610.

Rouse, W. B. (1991). *Design for success: A human-centered approach to designing successful products and systems*. New York: Wiley.

Rouse, W. B. (1992). *Strategies for innovation: Creating successful products, systems and organizations*. New York: Wiley.

Rouse, W. B., Cody, W. J., and Boff, K. R. (1991). The human factors of system design: Understanding and enhancing the role of human factors in system design. *International Journal of Human Factors in Manufacturing, 1*, 87–104.

Rouse, W. B., and Hammer, J. M. (1991). "Assessing the impact of modeling limits on intelligent systems." *IEEE Transactions on Systems, Man and Cybernetics, 21*, 1549–1559.

Rouse, W. B., Hammer, J. M., and Lewis, C. M. (1989). On capturing human skills and knowledge: Algorithmic approaches to model identification. *IEEE Transactions on Systems, Man and Cybernetics, 19*, 558–573.

Rouse, W. B., and Morris, N. M. (1986). On looking into the black box: Prospects and limits in the search for mental models. *Psychological Bulletin, 100*, 349–363.

Rumelhart, D. E. (1967). *The effects of interpretation intervals on performance in a continuous paired–associate task* (Tech. Rept. 116). Stanford, CA: Institute for Mathematical Studies in Social Sciences, Stanford University.

Scarr, S. (1985). Constructing psychology: Making facts and fables of our times. *American Psychologist, 40*, 499–512.

Sheldrake, R. (1981). *A new science of life: The hypothesis of formative causation*. London: Blond and Briggs.

Sheldrake, R. (1986). Morphogenetic fields: Nature's habits. In R. Weber (Ed.), *Dialogues with scientists sages: The search for unity* (Chapter 4). New York: Routledge and Kegan Paul.

Simon, H. A. (1975). Functional equivalence of problem solving skills. *Cognitive Psychology, 7*, 268–288.

Simon, H. A. (1991). *Models of my life*. New York: Basic Books.

Skinner, B. F. (1938). *The behavior of organisms*. New York: Appleton-Century Crofts.

Snow, C. P. (1965). *The two cultures: And a second look*. Cambridge, UK: Cambridge University Press.

Townsend, J. T. (1974). Issues and models concerning the processing of a finite number of inputs. In B. H. Kantowitz (Ed.), *Human information processing of a finite number of inputs* (pp. 133–185). Hillsdale, NJ: Erlbaum.

Voelcker, H. B. (1988). Modeling in the design process. In W. D. Compton (Ed.), *Design and analysis of integrated manufacturing systems* (pp. 167–199). Washington, DC: National Academy Press.

Watson, J. B. (1914). *Behavior*. New York: Henry Holt.

Weber, R. (1986). *Dialogues with scientists and sages: The search for unity*. New York: Routledge and Kegan Paul.

Whitehead, A. N. (1925). *Science and the modern world*. Cambridge, UK: Cambridge University Press.

Wright, R. (1988). *Three scientists and their gods: Looking for meaning in an age of information*. New York: Times Books.

Wulff, D. M. (1991). *Psychology of religion: Classic and contemporary views*. New York: Wiley.

Yarmolinsky, M. (1987). On impossibility in biology. In P. J. Davis and D. Park (Eds.), *No way: The nature of the impossible* (pp. 29–43). New York: Freeman.

Ziman, J. (1968). *Public knowledge: The social dimension of science*. Cambridge, UK: Cambridge University Press.

Zukav, G. (1979). *The dancing Wu Li masters: An overview of the new physics*. New York: Morrow.

Chapter 5

Elaboration and Integration of the Model

Chapters 2–4 outlined the basis for the Needs–Beliefs–Perceptions Model and provided a wealth of knowledge to be integrated into the model. This chapter focuses on this integration, as well as elaboration of the NBP Model. In addition, an NBP Analysis Methodology is discussed that provides the basis for pursuing the four archetypical innovation problems in Chapters 6–9.

An issue that emerged repeatedly in earlier chapters, but was not addressed, is the possibility of causal relationships among needs and beliefs. Are beliefs adopted because they satisfy needs? Or, are needs dictated by beliefs? The answers, I think, are "yes" to both questions.

If, for example, someone has great difficulty securing sufficient food for themselves and their family, they might adopt beliefs in whatever means consistently assures that the needs for food are satisfied. As another illustration, if someone has a strong need to belong and be esteemed, they might adopt the beliefs associated with an organization that warmly embraces them in membership.

Once adopted, however, beliefs may dictate needs. For instance, once someone firmly believes in the scientific method, they tend to exhibit the need to employ this method. In general, once someone has come to believe strongly in a particular set of traditions, they tend to have needs to celebrate these traditions.

Thus, there appear to be dynamic relationships among needs and beliefs. Needs influence adoption of beliefs. However, once adopted, beliefs influence the nature of needs. This explanation is, of course, much too simple, as the complexities of Chapters 6–9 illustrate. However, in using the NBP Model and NBP Template, it is important to consider both possibilities for the direction of causality.

Another aspect of the relationships among needs and beliefs is the basis upon which these relationships are determined. Chapters 2–4 reviewed much of what is known from a scientific point of view. Also reviewed was much material that is more philosophical and historical than scientific. Clearly, the knowledge base is inadequate for completely elaborating the NBP Model using only empirically and scientifically derived relationships.

As noted in Chapter 1, my approach to this dilemma is straightforward. To the greatest extent possible, analyses are based on knowledge gained by scientific studies. As the base of such knowledge is inevitably insufficient, the second source is heuristics based on a wide variety of personal experiences. The third source is well-reasoned speculation. In all cases, especially when the basis is speculation, I acknowledge the source of the presumed relationships.

This chapter is intended to be very practical. The concern is with using the NBP Model to provide insights into innovation problems. More specifically, the goal is to illustrate how the NBP Model can be systematically employed to diagnose problems—both barriers and hurdles—when attempting to create change. This chapter provides the methods and tools with which the problems in Chapters 6–9 are pursued.

CATALYSTS FOR CHANGE

The NBP Model can be used to explain—or, at least, hypothesize explanations of—perceptions of viability, acceptability, and validity. Such explanations can provide insights into how to change perceptions. These insights can serve as catalysts for change in the sense that changes—innovations—can be enabled by understanding the basis of perceptions.

Figure 5.1 depicts the NBP Model, elaborated to include catalysts. The function of catalysts is to utilize understanding of relationships among needs, beliefs, and perceptions to influence the model elements labeled "information" and "nurture." In this way, perceptions can potentially be changed.

Modifying Information

Understanding of relationships among needs, beliefs, and perceptions can be used to affect information provided to people. The central notion is to modify information content to satisfy needs without conflicting with beliefs. This presumes that needs and beliefs are taken as givens, because it is usually very difficult to change needs and beliefs solely with information.

Information can be modified in a variety of ways. If people are concerned that a pending organizational change, for example, will negatively affect satisfaction of their needs, information can be provided that illustrates how this negative effect

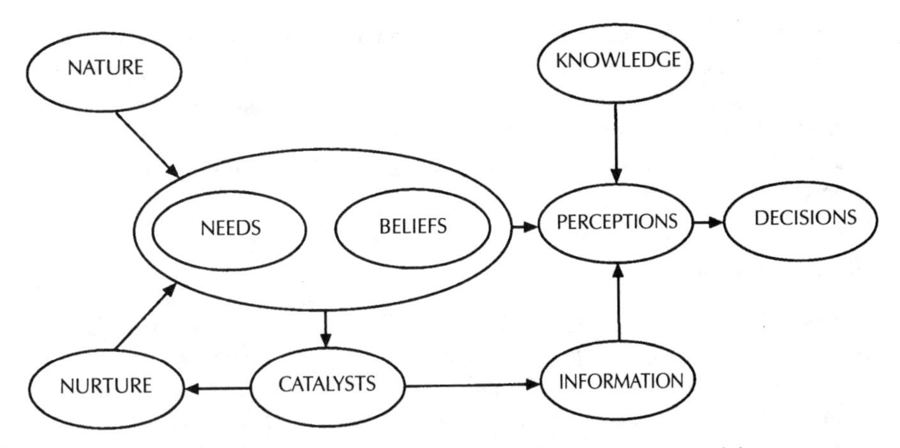

FIGURE 5.1. Elaborated Needs–Beliefs–Perceptions Model.

will not occur. As another illustration, if people perceive that a change will violate one of their closely held beliefs, they can be provided with information that shows this not to be the case.

This sounds much simpler than it actually is. As noted in previous chapters, people seldom express their negative perceptions in terms of unsatisfied needs and violated beliefs. Instead, they complain about surface features of the change—for example, practical reasons why it will never work. A needs–beliefs–perceptions analysis is often needed to uncover the underlying sources of the complaints.

Another difficulty is that, in many cases, changes are such that people's negative perceptions are correct. For instance, the planned downsizing of the company will result in lost jobs for some and less security for many. If information is modified to say that this will not happen—which, after all, is what people want to hear—it is unlikely to have any positive effect and quite likely to make the situation worse.

One possibility is to only stress the "upside" and avoid or ignore the "downside," hoping perhaps that no one will notice or complain. This tactic often works. However, it does involve sizable risks of later problems being worse than the problems at hand when the "downside" emerges.

Another way to deal with correct negative perceptions of change is to modify the nature of the change. It is quite possible that unsatisfied needs and violated beliefs are unnecessary by-products of change. If this is the case, then it may be possible to rethink the change so that it still accomplishes the original objectives but does not yield the same by-products. Consequently, information can be honestly modified in a manner that changes negative perceptions to positive.

Finally, there are situations in which nothing can be done to eliminate negative perceptions. In these cases, a needs–beliefs–perceptions analysis will at least make it such that one can deal with the real problem rather than just a surface problem. My experience is that people who are negative about a planned change feel somewhat better if management, for example, truly appreciates why they are negative. This is particularly true when the change is such that there is not much choice.

Affecting Nurture

Understanding of relationships among needs, beliefs, and perceptions can also be used to affect the way that people are nurtured, which eventually may lead to modifications of their needs, beliefs, and perceptions. This might be accomplished, for instance, via education, training, housing programs, or changes of work situations. Such mechanisms can lead to satisfaction of basic needs, emphasis of higher-level needs, and eventual adoption of new beliefs. With enough time, perhaps a generation or two, old beliefs may recede and the newer beliefs become dominant.

This also sounds much simpler than it really is. Especially problematic is the time necessary to affect changed perceptions in this way. Needs and beliefs are very difficult to change quickly. It requires much more time than is often available, or at least perceived to be available. Staying in business or getting reelected seldom seems to provide the time necessary.

Another difficulty is determining how needs and beliefs should be modified. History is rife with examples of people having no difficulty making this choice and frequently using force to assure that needs and beliefs changed accordingly, at least on the surface. We certainly do not want to emulate such examples.

However, we often unconsciously impose our belief systems on others. My experiences in developing countries in Africa, Asia, and South America have shown me how easy it is to feel, perhaps unconsciously, that you would like to help people in these areas to be just like us! While they may want the material aspects of our way of life, they do not necessarily want the same set of values and beliefs. In trying to help such people with the material side, we often impose an unspoken set of conditions on the values side.

Fortunately, many practical problems are much simpler. For instance, I have frequently been involved in situations where the primary change of beliefs required concerned the importance of customers. In particular, companies that try to become much more market oriented often discover that a sizable impediment is the attitudes of employees toward customers.

I have often heard statements such as, "Our product is technically excellent, the best in the market. If customers do not recognize this, it's their loss." A similar statement is, "Our customers are too dumb to really understand how innovative the system is that we are building for them." These statements reflect beliefs about customers that many companies would like to change. While it is difficult for a company to convince people that changes must occur, such an attitude adjustment typically is in the best interest of all employees and the company.

Specific examples of how to affect nurture are illustrated in Chapters 6–9. Perhaps the most important principle underlying this mechanism for catalyzing change is the simple recognition that nurture is affected by whatever you do—it cannot be avoided. Therefore, needs and beliefs will inherently evolve. Whether or not they evolve in a manner that is beneficial is highly related to the amount of attention that is paid to this process.

Summary

Catalysts for change involve two mechanisms. One has to do with modifying information so as to satisfy needs and not conflict with beliefs. This mechanism can affect perceptions relatively quickly. The second mechanism concerns affecting nurture so as to change needs, beliefs, and perceptions, albeit usually quite slowly.

If we consider these two mechanisms in the context of a vehicle metaphor, which provides the basis for the model of enterprise discussed in *Strategies for Innovation* (Rouse, 1992), the information mechanism can be viewed as a relatively fast "inner-loop" control process. In contrast, the nurture mechanism can be viewed as a relatively slow "outer-loop" control process. Inner loops tend to keep you on the road, while outer loops tend to assure that you reach desirable destinations. Coordination of both types of control loops helps to balance short-term and long-term objectives.

This type of coordination is also valuable when dealing in the realm of needs, beliefs, and perceptions. The information mechanism can assure that changes remain "on track." The nurture mechanism, however, provides the means for lasting change and long-term growth.

ANALYSIS METHODOLOGY

The utility of the NBP Model can be greatly enhanced if it is employed in the context of a methodology. While it is not necessary that a methodology be employed, its availability can help when it is not clear what to do next. In those situations where the next step is very clear, one can proceed without sticking to the methodology. Thus, a methodology can be of great help without also being an unnecessary constraint.

Figure 5.2 summarizes the Needs–Beliefs–Perceptions Analysis Methodology. This section describes this methodology in terms of the what, how, and why of

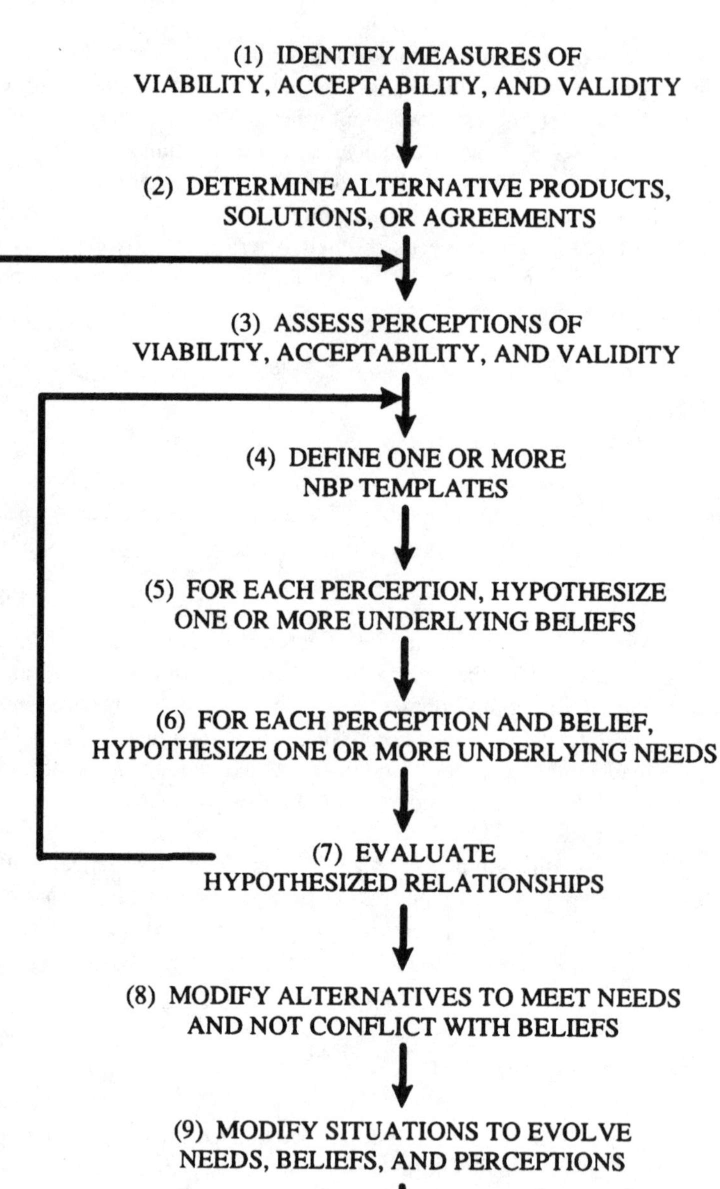

FIGURE 5.2. NBP Analysis Methodology.

each of the 10 steps. The actual performance of these steps is illustrated in detail in Chapters 6–9.

Identify Measures

Regardless of the type of problem and its context, the first step always involves identifying the stakeholders in the solution of the problem. For each class of stakeholder, measures of viability, acceptability, and validity are then identified. Methods and tools for accomplishing this step are discussed in *Design for Success* (Rouse, 1991) and *Strategies for Innovation* (Rouse, 1992).

Put simply, this step involves determining, for each class of stakeholder, the measures whereby they will judge whether or not a product, solution, or agreement solves the problem for them, solves it in an acceptable way, and solves it at an acceptable cost. This determines the ways in which perceptions are defined for each class of stakeholder.

Determine Alternatives

Alternatives tend to be very context specific. For our first archetypical problem—understanding the marketplace—we are interested in alternative products, systems, and services. For the problem of enabling the organization, the alternatives may be organizational structures, approaches to organizational change, educational methods, and so on. Alternatives in the context of sociotechnical disputes and political conflicts are usually possible resolutions of the dispute or conflict.

In many cases, the alternatives are known. There may be only one and it may be the source of dispute or conflict. It is quite likely that the NBP analysis will lead to new or substantially modified alternatives. Thus, it is not necessary to assure that all of the alternatives are defined at this point.

Assess Perceptions

This step involves assessing each alternative emerging from the second step in terms of the measures of viability, acceptability, and validity identified in the first step. This can be accomplished using questionnaires and interviews as is illustrated in *Design for Success* (Rouse, 1991) and *Strategies for Innovation* (Rouse, 1992). It also can be done more informally. However, less formal approaches tend to yield results that may be difficult to defend.

It is important that these perceptions be assessed, or inferred, in terms of the measures of viability, acceptability, and validity, instead of just overall perceptions. The finer-grained the measures, the easier it is to uncover underlying needs and beliefs. On the other hand, if the assessment is too detailed, people may lose patience with the process. The examples in Chapters 6–9 illustrate how this issue can be addressed.

Define NBP Templates

The results of assessing perceptions are compiled in one or more Needs–Beliefs–Perceptions Templates. As shown in Figure 5.3, and demonstrated in Chapters 3–4, this involves categorizing perceptions as positive or negative. This provides the starting point for determining the needs and beliefs likely to underlie these perceptions.

One NBP Template should be developed for each position on the issue at hand, or one for each class of stakeholder. For simple problems, such as illustrated in Chapter 4, it may be possible to include both pro and con positions on a single template. However, realistically complex problems typically require more than one.

Hypothesize Beliefs

As shown in the "beliefs" column of Figure 5.3, this step involves using Figures 3.6, 3.11, and 4.3 as a starting point for hypothesizing the beliefs likely to underlie the perceptions noted in the NBP Templates. This process is usually more complicated than simply selecting entries from these figures. Instead, these entries are used to prompt the analyst's thinking. As a result, and as illustrated in the examples in Chapters 3–4, the beliefs entered into the NBP Template are usually expressed in the context of the problem at hand rather than in the terminology of the referenced figures.

Initially, it may be necessary for an analyst to review portions of this book, or other materials, to be able to infer beliefs confidently. However, with practice in using this NBP Methodology, one quickly comes to recognize patterns that he or she has seen before. Consequently, this step is not as difficult as might be imagined, at least not once some experience is gained.

Hypothesize Needs

This step involves considering each pair of perceptions and beliefs. For each of these relationships, possible underlying needs are hypothesized using the information in Figures 3.5, 3.7, 3.10, and 4.2 as a starting point. As with beliefs, this process involves much interpretation and tailoring to the context at hand. However, I have found that this interpretation and tailoring tend to be a bit easier for needs than they are for beliefs.

It is quite possible for the same needs to underlie very different beliefs. For example, the need to belong can underlie membership in two quite different organiza-

ATTRIBUTE	PERCEPTIONS	BELIEFS	NEEDS
Viability	Positive	Entries Based On Figures 3.6 3.11 4.3	Entries Based On Figures 3.5 3.7 3.10 4.2
	Negative		
Acceptability	Positive		
	Negative		
Validity	Positive		
	Negative		

FIGURE 5.3. Needs–Beliefs–Perceptions Template.

tions with radically opposed belief systems. As a consequence, two different people may espouse very different beliefs despite common needs. The identification of these common needs may, however, provide a basis for creative solutions.

Evaluate Hypotheses

In this step, the hypothesized relationships among needs, beliefs, and perceptions are evaluated. One approach to this evaluation is to modify alternatives, or at least descriptions of alternatives, in ways that the hypothesized relationships predict desired changes of perceptions. Actual perceptions are then assessed and predictions confirmed or not.

It is important to note that results that confirm the predictions based on the hypothesized relationships do not prove the correctness of the relationships. Such results provide evidence of the *predictive* value of these relationships—which may be all that is needed—but do not prove the underlying constructs to be correct. If one is concerned with the *construct* value of the underlying relationships, a more systematic in-depth study is usually needed.

Modify Alternatives

This step involves considering the first mechanism discussed in the earlier section on catalysts for change, namely, modifying information. The basic idea is to modify information, as well as the alternatives if necessary, so as to meet needs and not conflict with beliefs. The goal is to change negative perceptions of viability, acceptability, and validity.

As was illustrated in Chapters 3–4, and is discussed in great detail in Chapters 6–9, the ways in which alternatives and/or information are modified are not always immediately obvious. Fortunately, for most problems, it is usually possible to discuss potential modifications with the stakeholders, or at least a subset of the stakeholders.

Modify Situations

In this step, the second mechanism discussed earlier is considered. This involves modifying situations so as to cause constructive evolution of needs, beliefs, and perceptions. As noted earlier, this might involve education and training, housing programs, and/or efforts to accelerate job creation.

This step, as well as the previous step, also involves considering how short-term and long-term strategies should be balanced. While it is often necessary to meet immediate needs, focusing solely on these needs may be counterproductive in the long term. Consequently, it is important to consider how situations might be changed, albeit usually slowly, to shift the nature of needs and beliefs.

Plan Life Cycle

The likely success of a mix of short-term and long-term strategies will be enhanced by having a cogent plan. This plan should include step-by-step implementation tasks and schedule. Also important are definitions of measures to be taken or events to be observed that will indicate success of the plan. There should be means for regular review of plans and results, as well as modifications as necessary.

Planning of this type is discussed at length in *Design for Success* (Rouse, 1991) and *Strategies for Innovation* (Rouse, 1992). As these references indicate, planning need not be overwhelming. However, the lack of plans, particularly for the types of problems discussed in Chapters 6–9, can result in overwhelming situations as, for example, initially positive perceptions unravel and there is no plan for dealing with such events.

Summary

This section has provided a brief overview of the NBP Analysis Methodology. While this methodology may appear overly structured, it should be remembered that the methodology only provides a *nominal* path for analyzing needs, beliefs, and perceptions. Those with experience and perhaps context-specific insights should not hesitate to approach this methodology much more eclectically than portrayed in Figure 5.2.

It is also important to note that use of this methodology can involve many subtleties. Rather than attempt to catalog all of these nuances, it is more instructive to discuss extensive case studies, as is done in Chapters 6–9. The next section provides an overview of these case studies.

OVERVIEW OF APPLICATIONS

The remainder of this book is devoted to applying the NBP Model, Template, and Analysis Methodology to four types of case studies:

- Understanding the marketplace,
- Enabling the enterprise,
- Settling sociotechnical disputes, and
- Resolving political conflicts.

These four archetypical innovation problems were introduced in Chapter 1 and noted in many of the discussions in Chapters 2–4. They are now the focus of our attention.

All four problems are concerned with change, the essence of innovation. Understanding the marketplace involves designing, developing, manufacturing, marketing, selling, and servicing innovative products, systems, and services in the global arena. The primary concern in discussing this problem (in Chapter 6) is what causes people to perceive—or not perceive—products, systems, and services as viable, acceptable, and valid. Put more simply, what causes people to accept an innovation and, consequently, change?

Enabling the enterprise is concerned with organizational change in general, and change in response to rapidly evolving market conditions in particular. The emphasis in this discussion (in Chapter 7) is on people's perceptions of planned changes of mission, markets, organizational structure, and their roles and jobs. What causes people to perceive—or, again, not perceive—that planned changes are viable, acceptable, and valid?

While the first two archetypical innovation problems deal with what can be easily viewed as win–win situations, the last two problems are concerned with what, on the surface, are typically viewed as win–lose situations. The premise of the discussions of these last two problems is that below the surface it may be possible to find win–win situations in terms of needs and beliefs. Of course, I cannot guarantee a silver bullet, even below the surface. However, a possible approach to dealing with apparently stalemated situations is certainly worth consideration.

Settling sociotechnical disputes involves dealing with multiple perspectives, typically conflicting perspectives. The nature of one or more measures of viability, acceptability, and validity is usually disputed. The assessment of alternatives relative to these criteria is also often hotly debated. Central to understanding such disputes is determining the needs and beliefs underlying the positions taken. Discussions (in Chapter 8) of how this understanding can be framed leads to the possibility of approaching disputes on several levels.

Resolving political conflicts often involves addressing long-held grievances. Priorities and justifications of claims and counterclaims usually dominate. Needs and beliefs of "haves" and "have-nots" strongly influence claims and demands. Discussion (in Chapter 9) of such conflicts emphasizes the possibility of dealing with issues that are deeper than the apparent, surface problems.

These four problems are addressed in their respective chapters using the NBP Analysis Methodology. The results of discussions in Chapters 2–4 are used as a basis for employing the methodology. Our understanding of needs, beliefs, and perceptions is drawn upon as shown in Figure 5.4. The entries in these general lists of needs and beliefs are augmented by additional sources targeted at the specific innovation problem at hand.

The discussions of the four archetypical innovation problems have several goals. First and foremost, the goal is to illustrate the benefits of needs–beliefs–perceptions analyses. These benefits include a structured process that enables

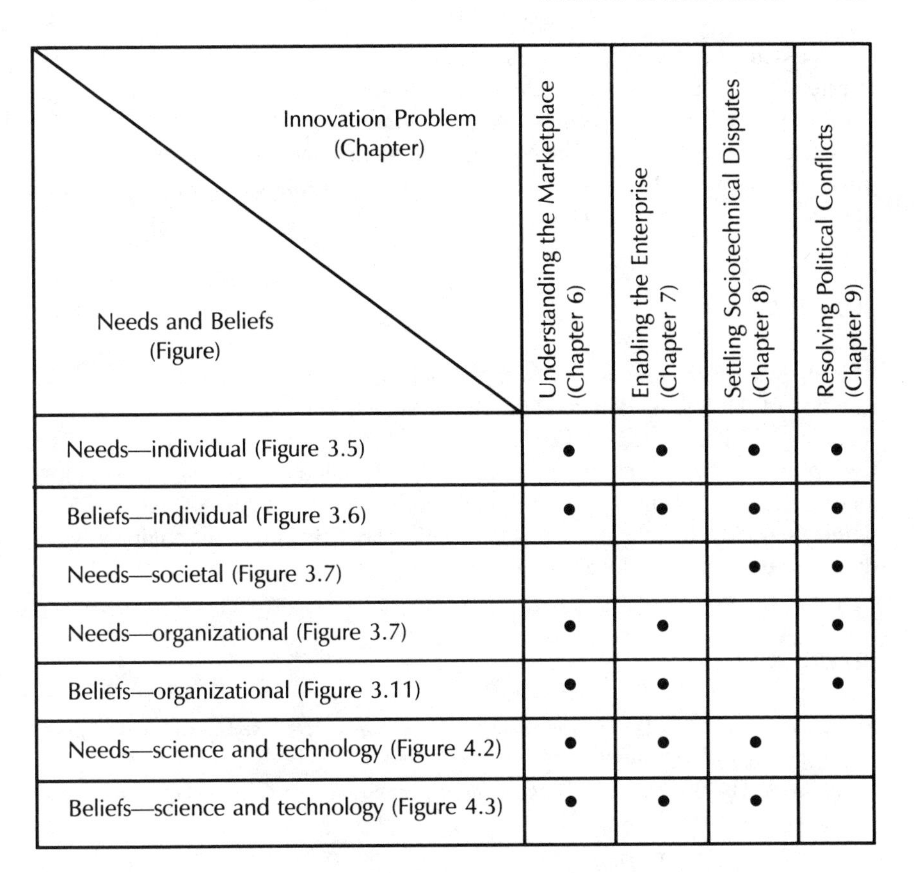

Needs and Beliefs (Figure) \ Innovation Problem (Chapter)	Understanding the Marketplace (Chapter 6)	Enabling the Enterprise (Chapter 7)	Settling Sociotechnical Disputes (Chapter 8)	Resolving Political Conflicts (Chapter 9)
Needs—individual (Figure 3.5)	•	•	•	•
Beliefs—individual (Figure 3.6)	•	•	•	•
Needs—societal (Figure 3.7)			•	•
Needs—organizational (Figure 3.7)	•	•		•
Beliefs—organizational (Figure 3.11)	•	•		•
Needs—science and technology (Figure 4.2)	•	•	•	
Beliefs—science and technology (Figure 4.3)	•	•	•	

FIGURE 5.4 Mapping of needs and beliefs to problems.

uncovering the underlying basis for perceptions and positions. This process helps to elicit and organize insights into why particular perceptions and positions emerge, as well as how they might be changed.

A second goal is to suggest creative alternatives for dealing with the four problems. For the first two problems—understanding the marketplace and enabling the enterprise—these alternatives are based, in part, on numerous practical experiences with addressing these problems. For the last two problems—settling sociotechnical disputes and resolving political conflicts—the alternatives are heavily influenced by personal experiences but, nevertheless, are a bit speculative. Hopefully, readers will find well-reasoned speculations to be somewhat intriguing.

A third goal of these discussions is to set the stage for planning and development of alternative solutions. Planning in terms of products, systems, and services is discussed in *Design for Success* (Rouse, 1991). Planning relative to an enterprise's mission, goals, and strategies is considered in *Strategies for Innovation* (Rouse, 1992). Chapter 10 brings the discussion of catalysts for change full circle back to these earlier books. The concepts, principles, methods, and tools in these books are shown to be applicable and useful in a much deeper way due to our understanding catalysts for change.

SUMMARY

This chapter has integrated the discussions and results of the first half of this book in terms of the Needs–Beliefs–Perceptions Model, Template, and Analysis Methodology. This included consideration of the ways in which the earlier compilations of needs and beliefs can be employed. While our knowledge of relationships among needs and beliefs is further augmented as discussions in Chapters 6–9 unfold, sufficient foundation is now in place to seriously pursue the complexity of the four archetypical innovation problems.

REFERENCES

Rouse, W. B. (1991). *Design for success: A human-centered approach to designing successful products and systems.* New York: Wiley.

Rouse, W. B. (1992). *Strategies for innovation: Creating successful products, systems and organizations.* New York: Wiley.

Chapter 6

Understanding the Marketplace

Trying to understand the increasingly global marketplace has become a much more difficult task. There was a time, I am told, in which people would buy almost anything. When I was a boy, people who owned Chevrolets always chose Chevrolets as their next car, a new one if they could afford it. People were similarly loyal to Fords and Plymouths.

The situation is very different now. There are a wealth of alternatives, purveyed by a wide range of sellers. Further, buyers seem much smarter now, not about everything, but certainly about shopping. Finally, technology has resulted in the dramatic shortening of the life cycle of many, but not all, products. While you can still sell basically the same airplane for 20 years, the selling life for consumer electronics is roughly two years.

To continue to prosper, companies have to be able to "delight" the marketplace with a steady stream of high-quality products. These products cannot be just slightly better versions of the same products sold for years. Instead, the marketplace expects innovations, expects change. Successful companies are those that can determine what the marketplace will perceive and embrace as innovative.

This concern quickly leads to the question, "What does the marketplace want?" Not surprisingly, the answer is, "It all depends." Different market segments emphasize one or more attributes such as performance, packaging, service, image, cost, and so on. Understanding market segments in terms of these attributes can help in the design and development of products, systems, or services.

This leads to the question, "What product, system, or service will delight the marketplace?" An approach to answering this question is to consider what is currently popular. A company might attempt to develop a way of providing higher-quality and/or lower-cost versions of current best-sellers. This would involve *process innovations* that enable improving quality and decreasing costs of production.

Alternatively, a company might attempt to satisfy an existing need in a new way. For example, rather than trying to compete with overnight express mail, one could focus on the use of facsimile transmission or electronic mail. While it is a little late to attempt to be a pioneer in this area, this example illustrates what are called *product innovations*.

Another type of product innovation involves meeting new needs, perhaps needs that are created by the availability of the new product, system, or service—examples include television and home computers. This is a risky, but potentially very high payoff, approach to innovation. The risk, of course, is that you will not succeed in anticipating and/or creating the needs necessary for the new market offering to prosper.

This risk is especially high for what I call "value-laden" products. Examples include springwater, soy meat, recycled paper products, and artificial fur. The success of these products depends on convincing potential customers of the importance of the values associated, usually very explicitly, with the products. These values include the importance of preventative health care, environmental protection, and animal rights for the examples cited. If potential customers do not perceive these values as important, they are unlikely to pay any premium for such products.

For all of the above types of innovation, a key to success is understanding the marketplace. In this chapter, the Needs–Beliefs–Perceptions (NBP) Model, Template, and Analysis Methodology are employed to illustrate how such understanding can be gained, particularly below the surface. This is accomplished in the context of two examples. One example focuses on marketing and sales of software products, with emphasis on a family of products that I have helped to introduce to the market.

The second example is concerned with transitioning defense technology to civilian applications. This case is based on an amalgam of numerous defense companies that I have attempted to help in making, or at least contemplating, this transition. These companies range from several small enterprises such as my own to large conglomerates that have been perennial leaders among defense contractors.

For both examples, the NBP Analysis Methodology is employed. The steps involving use of the NBP Model and Template are discussed in most detail. The other steps in the methodology, such as, planning life cycles, are outlined. Detailed consideration of these steps depends, to a great extent, on material in *Design for Success* (Rouse, 1991) and *Strategies for Innovation* (Rouse, 1992). Consequently, links to specific chapters in these other books are provided for readers who might like to pursue planning, for instance, in more depth.

MARKETING AND SALES OF SOFTWARE PRODUCTS

Since Search Technology was founded in 1980, we have produced a variety of software products and systems, ranging from computer-based training simulators to software tools for designing human–system interfaces. All of these products and systems were what is called "one offs," meaning we produced one, delivered it, and moved on to the next project. We never sold the same thing twice, although occasionally our customers resold the software many times to their customers.

We regularly discussed creating software products that we could sell repeatedly, with minimal or no tailoring for each customer. However, we seemed unable to prepare a coherent plan for accomplishing this goal. In retrospect, I think the primary problem was that we did not have a compelling vision of what particular products would delight the marketplace. We could imagine many products, but none of them seemed compelling.

This situation changed in 1989 when we launched the workshop series that eventually produced my book *Design for Success* (DFS), and in 1990 when we added the workshop that later yielded *Strategies for Innovation* (SFI). Initially, the workshops were primarily offered to our traditional customers. However, word of mouth quickly led to many new customers in a wide variety of enterprises—see Figure 1.4 for a summary.

We envisioned that the concepts, principles, and methods in these workshops and books would eventually form the basis for a set of software tools. However, we had not established a plan for developing these products. When we shared with customers our ideas for how these products might function, the response was overwhelmingly positive. A few customers almost demanded that we get started as they wanted these tools as soon as possible.

The result has been a family of products called Advisors. An Advisor is a computational framework that embodies a methodology, supported by explanation and tutoring functions that include expert systems and other types of aiding. The initial release—termed a beta release—of the SFI Advisor occurred in the fall of 1991, with two subsequent releases in 1992. The beta release of the DFS Advisor was in the spring of 1992, with a subsequent release later in the year.

This section presents an NBP analysis for these software products. This analysis focuses on understanding the needs and beliefs underlying perceptions of the SFI and DFS Advisors, and determining how this understanding should affect marketing and sales strategies. An additional goal, of course, is to illustrate the general applicability of the NBP Model, Template, and Analysis Methodology for marketing and sales of many types of products.

Identify Measures

The first step of the NBP Analysis Methodology concerns identifying measures of viability, acceptability, and validity. For the Advisors, measures are needed relative to tools for strategic planning (e.g., SFI Advisor) and product planning (e.g., DFS Advisor). These measures should be defined in a manner that is also applicable to methods and tools that compete with the Advisors.

Viability is related to benefits and costs. For planning tools, the benefits are concerned with people's perceptions of the value of planning and the value of the resulting plans. Costs relate to the time and effort required to become proficient with a method or tool, as well as the time and effort required to use a method or tool. There are also costs associated with acquiring and supporting a method or tool.

Acceptability concerns the willingness of individuals and organizations to adopt a product. People's perceptions of acceptability relate to whether a method or tool will "fit in," both organizationally and technologically. People are also often concerned with whether they are capable, both organizationally and culturally, of successfully using a method or tool. Another concern is whether management will support acquisition and use of a method or tool.

Validity relates to people's perceptions of the ability of a solution to solve their problems. There are two aspects to this issue. First, people must perceive that they have the problem that a particular method or tool is intended to solve. Second, they must perceive that this method or tool does, in fact, solve the problem. This second concern can be resolved by using the method or tool, seeing a compelling demonstration, or learning about other satisfied users.

Determine Alternatives

Our Advisors are clearly the alternative that we want customers to perceive as most viable, acceptable, and valid. However, customers have other alternatives that they will consider and, consequently, we must consider.

Many large enterprises already have product planning and strategic planning methods and tools. Fortunately—for us—it is frequently the case that people find existing methods overwhelming, yielding much paper but little value. Unfortunately, however, these methods and tools are incumbents which we have to unseat. This can be very difficult.

Most small enterprises, and a surprising number of large enterprises, have no methods and tools. They also do little or no planning. If they say that they do planning, they typically mean budgeting and setting next year's sales targets to satisfy budget requirements. These types of customers often have to be convinced that planning will be of value.

There are also other planning methods and tools in the market. However, thus far, we have found that the primary alternatives against which we compete are established corporate planning processes, even if they are disliked, and a lack of planning processes and apparent lack of perceived value of planning. These two alternatives represent two different types of status quo.

Assess Perceptions

While measurement is planned by first considering viability, then acceptability, and then validity, measurements are actually taken in the opposite order—see Chapter 2 of *Design for Success* (Rouse, 1991) for detailed discussion of this process. This is due to the simple fact that customers are first concerned with whether a solution can solve their problem. They subsequently concern themselves with whether the solution is acceptable and whether benefits exceed costs.

Considering validity, therefore, the first issue noted earlier was the extent to which potential customers perceived that they had planning problems. Most large enterprises easily admit to having such problems. They try to plan, but do not seem to produce good plans.

For small- to mid-sized enterprises, planning is less prevalent. Top management often perceives that it should start to plan in terms of mission, goals, and strategies. It also perceives that it should plan for the evolution of its products, systems, and services. However, it often is not sure that it is ready for formal planning processes.

The second aspect of validity concerns customers' perceptions of the appropriateness of a particular method or tool. As emphasized in *Strategies for Innovation* (Rouse, 1992), the Advisors provide an integration of quantitative and qualitative characteristics of plans. While financial considerations are very important, they do not dominate plans created using the Advisors. For potential customers who viewed planning in general as synonymous with financial planning, initial releases of the Advisors tended to seem weak on the financial side.

If a method or tool is perceived to be valid, it will not necessarily be perceived as acceptable. In a few cases, potential customers perceived the Advisors to be valid but not acceptable because these tools did not reflect the peculiarities of their planning processes. These types of customers typically said that if they were "starting from scratch," they would adopt the processes embodied in the Advisors.

Another common acceptability concern expressed by potential customers was their perception that they were not organizationally capable of becoming proficient in planning. This took two forms, one involving self-doubt and the other concerned with skepticism about their top management. These types of enterprises usually want substantial support in changing their organizational processes.

A third acceptability issue related to technological compatibility. The initial releases of these products were targeted at the largest installed base of computers. We anticipated eventually offering versions of this software for other computers, but not until the initial releases had established themselves. This presented difficulties for potential customers whose existing computers—in two cases, computers that they produce—were not compatible with the computer that we targeted with the initial releases. Those who judged the Advisors high in terms of validity said, in most cases, they would not give us high marks for acceptability until the versions they needed were available.

Perceptions of viability centered on people's perceptions of the difficulty of planning (i.e., the cost of doing it) and their perceptions of the value of the likely results (i.e., the benefits of having plans). Many large organizations have experienced planning to be very time consuming, with little payoff. Most small enterprises, on the other hand, have not had to plan extensively before and, therefore, are not sure of how difficult it will be and what benefits will result.

Interestingly, we found that perceptions of viability were only weakly related to the purchase price of the software. We priced the Advisors so that customers could start with a fairly modest purchase and expand to more comprehensive planning support once they were confident of the approach and their abilities to succeed with it. Potential customers were much more concerned with the costs that they would incur *after* they bought the Advisors, particularly in terms of their time.

Just as differences were noted between large and small enterprises' perceptions of validity, acceptability, and viability, differences were found between corporate and government perceptions, and between United States and foreign perceptions. While considerable enthusiasm was encountered with several government customers—especially for workshops—such customers found it exceedingly difficult to adopt new planning processes. The problems were mainly acceptability and viability. The Advisors were viewed as very different and perhaps not compatible with existing bureaucratic processes. They also suffered from customers' overall skepticism that serious planning is of value in highly politicized environments. While we experienced some success in overcoming these negative perceptions, these few successes can be attributed to a limited number of very committed people.

In considering the differences between United States and foreign perceptions, the primary difficulty encountered was language differences. While people might be able to accept an Advisor's explanations and tutoring in English, they perceived that they might not be able to express their thoughts—especially strategic thoughts—in other than their native language. Fortunately, there are a variety of character sets, that is, different alphabets, available that enable users to respond to the software's prompts in many different languages.

Overall, the differences between large and small enterprises, and the differences between government and industrial enterprises, were much greater than the differences between United States and foreign enterprises. This conclusion is, of course, likely to be dependent on the types of enterprises of interest. For technology-based enterprises, such as we have pursued as customers, it may be that the international nature of high technology results in many cross-cultural differences being minimized.

Define NBP Templates

Use of the NBP Templates could be approached in several ways for this problem. For any particular potential customer, one template could be formed for the Advisors and another formed for the incumbent planning processes. If the analysis concerned a sample of all potential customers, one template could reflect the Advisors and one could be formed for each of the predominant alternatives.

For the purposes of this case study, only one NBP Template is needed. As shown in Figure 6.1, the positive perceptions noted are those that support adopting the Advisors, while the negative perceptions indicated a lack of support for adopting them. The negative side of this template reflects more than one alternative, including any incumbent planning processes and the possibility of no planning processes.

Hypothesize Beliefs

The beliefs indicated in Figure 6.1 were drawn from two sources. First, they are based on the understanding of the problem elaborated in the earlier steps of the NBP Analysis Methodology. Second, these entries were reviewed and modified somewhat by using Figures 3.6, 3.11, and 4.3 as reference compilations.

Note that the wording in Figure 6.1 is specific to the context on planning and associated methods and tools. It is not necessary to retain the wording from the discussions in Chapters 2–5. Context-specific explanations are usually much more useful.

The beliefs associated with positive perceptions reflect the earlier discussion of why people find the Advisors to be valuable. Similarly, the beliefs associated with negative perceptions describe the aforementioned reasons for rejecting the Advisors. For this particular problem—adopting or rejecting new planning methods and tools—people tended to express their perceptions in terms of their beliefs. This degree of candor is an exception, as later examples illustrate.

Hypothesize Needs

In contrast to beliefs, people seldom express their perceptions directly in terms of needs. We must infer possible needs and then evaluate these inferences. Based on Figures 3.5, 3.7, 3.10, and 4.2, as well as understanding of the context of the problem, the underlying needs indicated in Figure 6.1 were hypothesized.

Considering all of the needs associated with positive perceptions, a pattern emerges. People supportive of the Advisors tend to be problem solvers who like structured approaches, want to understand and control the future, believe that they

ATTRIBUTE	PERCEPTIONS	BELIEFS	NEEDS
Viability	Positive	Planning is useful, plans will help, and costs of adoption are tolerable	Need for structure, need to be in control and solve problems, lack of need to avoid risk and failure
	Negative	Planning is too difficult, plans seldom are useful, or adoption will be too costly	Need to deal with fear of uncertainty and fear of failure
Acceptability	Positive	Approach fits organizationally and technologically, capability to employ approach is adequate, and adoption will be supported	Need for creativity and change, need to believe in making a difference
	Negative	Approach will not fit in, capability is limited, or adoption will not be supported	Need to fit in, need for consistency and continuity
Validity	Positive	Planning problem exists and approach will help	Need too reason things out, need for coherence and consistency, need to be in control and solve problems
	Negative	Planning problem does not exist or other, more financially oriented approaches, are more appropriate	Need to deal with fear of uncertainty and fear of failure, need to employ and exercise particular skills

FIGURE 6.1. NBP Template for marketing and sales of software products.

can make a difference, and are not overly concerned about uncertainty and failure. In contrast, people with negative perceptions tend to be stymied by uncertainty, want the future to be like the past, and want their own well-learned approaches to be the solution.

This explanation of the differences between supporters and detractors appears to classify supporters as smart and optimistic, and detractors as dumb and pessimistic. This is not what is intended. The supporters could be described as "change agents," regardless of their intelligence and level of optimism. The detractors, on the other hand, were more tied to the status quo, which in some cases was appropriate—they were doing quite well—and in many others was not a good stance. In some cases, the detractors were intelligent, optimistic people who simply disagreed with the idea of trying a new approach.

Evaluate Hypotheses

Hypotheses regarding beliefs and needs are usually difficult to evaluate in a rigorous experimental manner. Less formal, but more comprehensive and realistic, approaches involve using potential or actual customers for the product, system, or service of interest as the basis for evaluation. For the Advisors, we employed our initial set of users, as well as many more enterprises that had participated in the workshops associated with the Advisors.

This evaluation involved a careful analysis of our successes and failures, including a large set of potential customers where we had, thus far, failed to succeed. For this large set, success was still a possibility. Via numerous face-to-face discussions and a large number of telephone conversations, we were able to confirm our hypotheses. This confirmation was not rigorous in a scientific sense, but it was quite sufficient to enable us to employ our understanding of needs, beliefs, and perceptions as catalysts for changing the Advisors.

Modify Alternatives

This step concerns possibly modifying alternatives to better meet needs of detractors without conflicting with their beliefs. For the Advisors, we were concerned with detractors who felt that they had planning problems, but did not perceive the Advisors to be the solution. Our goal was to modify the Advisors so as to gain the support of the detractors without losing existing supporters.

To enhance perceived validity, the financial modeling capabilities of the Advisors were substantially enhanced. Thus, more financially oriented users found more of what they wanted. We also enhanced validity by creating export functions that enabled output of information created in the Advisors to, for example, more powerful financial modeling packages and desktop publishing packages.

To enhance perceived acceptability, the "setup" function within the Advisors was substantially extended. This function enabled users to tailor the Advisors in several ways. Spreadsheet format, labels, and equations can be tailored. Formats and wording in templates can be modified. Enterprise-specific processes can be represented and linked to elements of the processes in the Advisors. Enterprise-specific expert advice can be added.

Additional modifications to enhance perceived acceptability were the provision of limited, but free, remote support to answer questions and give advice. On-site consulting to support customers through the "learning curve" was also packaged with many proposals. While this raised the price, it also seemed to increase perceived viability because the costs of introduction and initial use were now fixed.

These changes resulted in many, but not all, detractors becoming supporters. More importantly, new potential customers, with characteristics similar to those who had been detractors, were now often supporters. The key to accomplishing this change was understanding the relationships among customers' needs and beliefs as they affected their perceptions of the validity, acceptability, and viability of the Advisors.

Modify Situations

Not all detractors or potential detractors were "converted." Especially troublesome have been those who believe that financial planning is synonymous with planning in general. These people, not surprisingly, tend to have a strong need to employ their financial skills. They often make statements such as, "It does not matter what the product and market are, as long as the finances look good."

This is 1980s thinking. "Managing by the numbers" is slowly, but surely, being discarded as the dominant paradigm in innovative companies. Of course, many people skilled in this waning paradigm still occupy positions of power.

In order to slowly, probably very slowly, change the "numbers only" needs and beliefs, we continue to offer workshops that emphasize the integration of non-financial and financial aspects of strategic thinking. We have also broadened the range of outlets in which articles and editorials are published, and presentations are made.

It is likely that it will take 5–10 years to change the orientation of top management of many enterprises. Fortunately, many people are contributing to heralding the necessary changes. From the perspective of selling Advisors, we have to provide products and services that are compatible with current needs and beliefs, and smoothly evolve to be compatible with future needs and beliefs.

Plan Life Cycle

If the Advisors are to remain innovative, they will have to continually change. Successful changes will have to anticipate the evolution of potential customers' needs and beliefs regarding planning itself, as well as planning methods and tools. Consequently, we have to plan for the life cycle of the Advisors, not just the next offering.

A key ingredient in this planning is anticipating how customers will change. One of the best ways to do this is to involve customers in the evolution of the product family. We have found that this results in numerous comments and suggestions for how current products can be improved. This process also provides many insights into customers' thinking and intentions.

This use of the marketplace as a sounding board can be done both systematically and efficiently, to yield the most insights in the least time. Much of *Design for Success* (Rouse, 1991) and *Strategies for Innovation* (Rouse, 1992) describes and illustrates how this can be done. The result of this process can be a rapid upgrading of products, incorporating market-driven functionality. In this way, you can eclipse your own products before your competition does.

Summary

This section has focused on marketing and sales of software products, with emphasis on a particular family of products. Using the NBP Analysis Methodology, the needs and beliefs underlying differing perceptions of the products were hypothesized and evaluated. The results of this analysis led to modifications of the products to yield positive perceptions by a broader range of potential customers.

TRANSITIONING DEFENSE TECHNOLOGIES TO CIVILIAN APPLICATIONS

United States defense expenditures (in real dollars) for procurement peaked in 1986 or 1987. Since that time, defense contractors have been competing for pieces of a pie that is getting smaller and smaller. The result has been fiercer competition for smaller profits and fewer jobs. Many thousands of defense workers have been laid off in the process.

There is a similar trend worldwide. Defense contractors in many, if not all, of the countries where we do business are scrambling for sales, and trying to use export sales to compensate for decreased domestic sales. Many of the participants in our workshops attend in hopes of learning how to succeed in a declining defense market.

I have found that the differences among defense contractors across countries are much smaller than the differences between defense and civilian markets. In other words, defense contractors are much more alike, regardless of country, than are defense and civilian companies within any particular country. Consequently, most defense companies around the world face similar difficulties in trying to transition to civilian markets.

I can illustrate some of these difficulties with two examples. In the late 1960s, I worked for a large defense contractor. As I was preparing to leave to go to graduate school, this company announced major plans for pursuing civilian markets. They correctly anticipated the decline in defense expenditures following Vietnam. As I was leaving, they had just dedicated a new building for this commercial venture.

A year or two later, I happened to meet a manager with whom I had worked at this company. I asked him how the transition was going. Somewhat facetiously, he commented, "It's really difficult to get a weekend boater to buy a $1 million depth finder!" I later learned that the company was having great difficulty getting nondefense sales above 1 percent of sales.

Although I do not know the details, my guess is that their depth finder was designed for high performance and was able to withstand vibrations from depth charges and artillery attacks. However, the civilian market did not value the level of performance achieved and did not need the degree of ruggedness built into the product. Moreover, of course, the price was much, much too high.

Just recently, I was talking with a potential customer whose defense business, and number of employees, had decreased by half. I asked him what his company was doing to compensate for the dwindling market. He said that it was inundating defense agencies with proposals. I asked if the company really thought this tactic would work. He said, "No, but it's the only thing that we know how to do."

Talking with executives in this company, it was clear they knew they should try to transition to nondefense markets. However, they were not sure of what such a transition meant. It was not clear what they should do. Furthermore, their defense business, what was left of it, was stable for the moment. Nevertheless, the CEO said to me, "Every morning, I read the paper to see if we are still in business."

In the first example, the company thought it knew what to do, but in retrospect did not. In the second example, the company recognized the problem, but did not know what to do. In both cases, the problem was not one of avoiding the need to transition. The problem was a lack of detailed knowledge of how to transition.

There are two aspects to answering the question of how to transition. The first concerns understanding the marketplace. How do defense and civilian markets differ? How do customers' perceptions in these two types of markets differ? Why are there differences in perceptions? In this section, the NBP Model, Template, and Analysis Methodology are used to answer these questions.

The discussions in this section yield an understanding of the marketplace. This provides a good starting point. However, we also must deal with the second aspect of this problem. How can a defense company be enabled to succeed in civilian markets? Answering this question involves considering the organizational changes necessary to respond to different markets' needs, beliefs, and perceptions. Chapter

7 pursues this second aspect of transitioning defense technologies to civilian applications.

Identify Measures

This step involves identifying measures of viability, acceptability, and validity. This requires that we identify whose perceptions are of interest. Since this chapter focuses on understanding the marketplace, the stakeholders of concern are customers and users. Other stakeholders, such as employees, management, and shareholders, are considered in Chapter 7.

We also are interested in differences between defense customers and users, and civilian customers and users. Understanding these differences is essential in determining how products, systems, and services must change. Thus, this analysis considers and contrasts these two groups.

In terms of viability, defense customers and users clearly focus on the benefits of high performance and often seek levels of performance not previously demonstrated. Cost concerns usually center on the difficulties and time required to integrate new products and systems into use. The cost of acquisition is not as big a concern—although, it is not disregarded—as the desire for high-tech, state-of-the-art performance capabilities.

Civilian customers and users of products and systems are much more conscious of cost-benefit trade-offs. High-performance and state-of-the-art technology are only valued to the extent that they produce "bottom-line" benefits in excess of their cost. In the United States, this payback is expected in 1–3 years, while in Japan and Europe 5–10 years may be acceptable.

As regards acceptability, defense customers and users have massive bureaucratic processes for regulating change. It is difficult to get a new product or system accepted because "fitting in," in my experience, is a long, arduous process. However, once your product or system has been accepted, you can sell it, as well as upgrades, to defense agencies around the world for decades.

In contrast, civilian customers and users accept and discard products, systems, and services much more quickly. As illustrated in the earlier example of software products, acceptability problems can arise in terms of organizational and technological compatibility. Further, other new products, systems, and services—if they can make the viability hurdles—are often also considered. Hence, getting accepted and staying accepted are much faster processes in civilian markets.

Assessing validity in defense markets often involves formal test and evaluation processes where performance is measured and compared to requirements. While there are often penalties for not passing tests, these penalties seldom result in a contractor no longer being a supplier of the product or system in question. Instead, products or systems are redesigned, or expectations are lowered so that success can be achieved. This tactic is necessary because defense agencies often, at this point in the development of a new product or system, no longer have a choice of alternative contractors. It is usually much too late.

From a procedural point of view, in some types of civilian markets, validity is assessed in much the same way, typically using different measures such as throughput or quality. However, the use of such measurements is somewhat different in these civilian markets. Failure to meet requirements usually involves the manufacturer incurring the costs of redesign or, more simply, losing sales and market share.

For other civilian markets—particularly consumer markets—validity is assessed quite differently. While consumer groups may test and evaluate products, the assessment process is usually much more informal. Consumers rely on endorsements by opinion leaders, comments of their peers, and their own past experiences. If, for example, an automobile manufacturer has a reputation for poor quality, its newer cars will be judged to be low in quality independent of their actual quality. Eventually, the company may rebound in consumers' perceptions, but it will happen much slower than the actual improvement of their products.

The differences among defense and civilian markets are, in my experience, very substantial. Defense customers and users focus on high performance with minimal concern for costs, very slowly adopt or discard products and systems, and tend to stay with contractors even if expectations are not met. Civilian customers and users focus on performance to the extent that bottom-line performance vs. cost trade-offs make sense, relatively quickly adopt or discard products and systems, and only tend to stay with vendors if expectations are met. Clearly, transitioning from defense to civilian markets is no easy matter.

Determine Alternatives

In this discussion, the concern is not with a specific product or system. Instead, the focus is on a particular problem, namely, applying technology developed for the military to designing and developing civilian products and systems. The easiest transition of this type involves using defense technology for customers such as the National Aeronautics and Space Administration (NASA) or the Federal Aviation Administration (FAA) in the United States, or equivalent agencies in other countries.

One reason for this is that one is still dealing with the government, a skill usually highly developed among defense contractors. Another reason is that, in my experience serving on advisory committees for NASA and FAA, these agencies are very similar to defense agencies. They are mission oriented, rather than customer or market oriented. In addition, they are often technology driven, rather than being driven by how they can best support their end users.

This alternative—nondefense government—is not considered further in this chapter. This is due to the fact that this market is much too small to be the primary alternative for most defense contractors. Most of these companies need, and the economy needs, to transition successfully to nongovernment applications of high technology, as well as applications of the expertise and experience underlying this technology.

To illustrate use of the NBP Analysis Methodology, I will employ as examples four types of nongovernment applications that we have attempted. In all cases, our goal was to transition software technology developed for military aerospace applications to civilian use. We pursued four potential domains of application: commercial aviation, discrete parts manufacturing, medical equipment, and sales/service support systems. Only the last of these four has been an unqualified success. The other three are slowly percolating, at best.

To explore the difficulties underlying these transitions, all of the potential customers and users in these domains can be viewed as a single type of customer and user, referred to as *civilian* customers and users. Similarly, all military customers are referred to as *defense* customers and users. In the remainder of this section, the needs, beliefs, and perceptions of these two classes of customers and users are contrasted.

Assess Perceptions

The first comparison of interest involves perceptions of validity. Defense markets often will perceive novel high-tech solutions, backed perhaps by analytical evaluations, as valid, or at least valid until proven invalid. Civilian markets, in contrast, are usually much more skeptical. Often, they want empirical proof that somebody's bottom line has been improved by employing the technology.

Proof of validity for civilian customers and users is not straightforward. The fact that "proof of concept" systems have been developed and rigorously evaluated for defense applications often provides no credibility whatsoever. In fact, in the United States at least, there is often skepticism about the abilities of the military to do anything right—while this assertion is difficult to justify, it nevertheless is a common perception. Another problem is that customers and users in one domain often reject evidence from other domains. Thus, for example, potential manufacturing customers want evidence from other manufacturing applications, not from aerospace applications.

In considering acceptability, defense customers tend to perceive products and systems to be acceptable long before they actually are accepted. As noted above, the military services have a long and formal process of integrating new products and systems into fielded use. Beyond this time-consuming process, however, they have relatively few acceptability problems because they pay all the costs of developing a product or system to meet their requirements.

Civilian customers and users, on the other hand, do not pay these costs and, therefore, have to choose among available products and systems. This can result in high-technology solutions encountering the NIH and NSH phenomena. NIH is the well-known "not invented here" syndrome, whereby a new solution is undermined by customers' in-house technical people who think that they have a better alternative.

NSH is the less well-known "not supported here" concern. Customers are usually very concerned with the life cycle of the product or system, not just the acquisition costs. They want to be sure that the product or system can be supported throughout its lifetime, in both an effective and economical manner. If the product or system embodies technologies, including processes, that none of the customer's in-house technical people understand, it is quite possible that NSH will be the result.

Another aspect of perceptions of acceptability concerns technological innovations. Defense markets often highly value such innovations—they keep us ahead of our adversaries. Civilian markets, in contrast, typically associate technological innovation with risk. Numerous civilian customers and users have told me that they try to avoid innovation, to the extent possible. To them, innovation means change, change means something new, something new means something not yet mature, and all of these things mean risk. Many civilian customers would rather have a well-worn solution that solves their problem than an innovative solution that does more than solve their problem.

I hasten to clarify this apparently categorical conclusion. I have found this phenomenon most prevalent in discrete parts manufacturing and commercial aviation. In industries such as computer hardware and software, as well as electronics, this phenomenon occurs much less frequently. My explanation is that highly competitive industries, with short product life cycles, have no choice but innovation. Otherwise, their products will be obsolete in two to three years.

As noted in earlier discussions, perceptions of viability also differ between defense and civilian markets. Defense customers typically want maximum performance, with acceptable life-cycle costs. Since the goal is to be able to perform better than adversaries, and it is seldom certain how well adversaries can perform, the emphasis is on using any technologies, sometimes at almost any cost, that will assure a performance advantage. The result can be, for instance, aircraft such as the B-2 which are projected to cost almost $1 billion each.

Civilian markets are much more cost sensitive. They only want as much performance as the market demands which, as illustrated earlier, does not usually require the maximum possible available with state-of-the-art technology. Civilian customers and users focus on the bottom line in two ways: revenues and costs. From a revenue perspective, concerns emphasize the marketplace's perceptions of the value of benefits potentially provided by technologies. Put simply, the revenue question involves estimating how many units can be sold, at what price, with and without the technologies of interest. The cost question concerns the life-cycle costs of employing the technologies.

Trade-offs between revenues and costs depend on the time horizon considered. Payback time, namely, the time needed to recoup investments, is important. Also critical is the expected revenue-generating life of the product or system since, after all, the goal is not simply to recoup investments. Given a time horizon, and a discount rate for future revenues and costs, this trade-off can be approached quite rigorously. These types of calculations are much more difficult for defense customers because defense—life, limb, and property—is difficult to account for in the same units as costs.

Define NBP Templates

To contrast the needs, beliefs, and perceptions of defense and civilian markets, two NBP Templates are used. One template, Figure 6.2, summarizes the needs and beliefs underlying perceptions of defense customers and users. The second, Figure 6.3, focuses on civilian customers and users. The discussion in the remainder of this section considers the differences between these two figures, and the implications of these differences for transitioning technology from defense to civilian applications.

It should be noted that these two figures are aggregations across agencies (Fig. 6.2) and market sectors (Fig. 6.3). This aggregation is reasonable for the purposes of illustration. For a real market analysis, however, the NBP Template should be focused on the specific market of interest. Consequently, a number of templates might be necessary to consider seriously all of the possibilities for transitioning particular technologies.

Hypothesize Beliefs

The beliefs columns of Figures 6.2 and 6.3 were filled in by first reviewing the results of the earlier steps in this analysis and then consulting Figures 3.6, 3.11, and 4.3 to prompt any additional insights. As noted in earlier analyses, no attempt was made to retain the wordings in the figures from earlier chapters.

If one compares beliefs across the defense and civilian markets, several strong differences are very clear. Defense customers and users believe in striving for the maximum possible performance in order to better enemies with uncertain capabilities. Innovation is highly valued, since it usually promises the greatest performance. Technology that meets requirements is judged as good—often, there is no other "bottom line."

In terms of the civilian market, understanding of the market drives requirements which seldom dictate maximum performance. The value of technology is assessed in terms of its impact on the bottom line, by reducing costs and/or increasing sales. Innovation is not, by any means, inherently good.

From an acceptability perspective, defense customers and users focus on meeting requirements. They dictate requirements, pay contractors to meet them, and

ATTRIBUTE	PERCEPTIONS	BELIEFS	NEEDS
Viability	Positive	Enemies' intentions and plans are key, maximum performance required, meeting requirements sufficient, innovation inherently good	Need for enemies, need for performance advantage
	Negative	Less than maximum performance not sufficient	Need for performance advantage, need to deal with uncertainty
Acceptability	Positive	Meeting requirements sufficient, contractors paid and forced to meet requirements, long product life cycles	Needs to be in control and judged well by others, fit into tradition, play by the rules, and avoid risks
	Negative	Off-the-shelf technology not good enough	Need for performance advantage
Validity	Positive	Newest technology is best, formal validation possible	Need to defend choices
	Negative	Mature technology provides no advantage	Need for performance advantage

FIGURE 6.2. NBP Template for defense customers and users.

force compliance if necessary. Long product life cycles are the norm, in part because off-the-shelf technology is avoided.

In contrast, civilian customers and users expect technology to be off-the-shelf and inherently compatible from technical, organizational, and cultural points of view. If modifications are necessary to assure compatibility, they do not expect to pay for these modifications. Short product life cycles are the norm, with new proven technology replacing older technology as soon as it can be justified.

ATTRIBUTE	PERCEPTIONS	BELIEFS	NEEDS
Viability	Positive	Market drives requirements, competitors' intentions and plans are key, bottom line is what counts	Need to respond to market and perform better than competitors, need to recoup investment and make profit
	Negative	Maximum performance seldom required, value and quality do not require high price	Need to respond to market
Acceptability	Positive	Off-the-shelf technology should be inherently compatible, should not pay for compatibility, short product life cycles	Need to be in control and judged well by others, need to act quickly, need to have own solution adopted
	Negative	Not invented here, not supported here, tailoring should not increase price	Need to avoid risk, need to support technology
Validity	Positive	Mature technology is best, reference sales in own domain are best evidence	Need to defend choices, need to have own solution adopted
	Negative	Newest technology, military technology, and innovation in general are risky, lack of domain-specific evidence is risky	Need to avoid risk

FIGURE 6.3. NBP Template for civilian customers and users.

Defense markets inherently value the newest technology. They pay for R&D as well as design, manufacturing, and support. A variety of formal "validation" exercises are imposed, although it is usually quite difficult to assess more than compliance with requirements.

Civilian markets believe in using mature rather than new technology. Maturity is defined by one or more success stories in related domains of application. Military applications are seldom viewed as relevant or credible. Innovation is considered to be risky.

Hypothesize Needs

The needs columns of Figures 6.2 and 6.3 were completed by reviewing the results of previous steps of this analysis, as well as the entries in Figures 3.5, 3.7, 3.10, and 4.2. As explained earlier, no attempt was made to retain the wordings from these figures.

Note that it is quite possible for the same needs to underlie multiple beliefs. Further, these need–belief relationships may cut across perceptions of viability, acceptability, and validity. Thus, as illustrated by Figures 6.2 and 6.3, needs and beliefs do not have to be crisply compartmentalized.

Several interesting contrasts emerge by comparing the needs columns of Figures 6.2 and 6.3. Defense customers and users need enemies. They tend to assume that their enemies are very capable. They need to have a performance advantage over these enemies.

Civilian customers and users, on the other hand, would rather not have competition. They try to respond directly to the desires of the marketplace—competitors serve to complicate the process of responding. The civilian market is heavily affected by needs to recoup investments and make profits. If technology does not contribute to meeting these needs, and especially if it serves to undermine them, it will be avoided.

Defense customers and users are heavily affected by the need to fit into traditions, which vary across services, and by the need to play by procurement rules. Civilian customers and users, on the other hand, need to act extremely quickly with minimal overhead and paperwork. Since civilian customers and users can develop their own solution, which the government seldom does, it is common for in-house solutions to be advocated. Competing external solutions are labeled NIH and NSH (again, "not invented here" and "not supported here," respectively).

The two markets are similar in several ways. Customers and users feel needs to be in control, judged well by others, and able to defend their choices. Risk avoidance is common in both markets, but relates to different concerns. Defense customers and users avoid being perceived as taking risks relative to the process of procurement and fielding of systems. They have little bottom-line risk—their jobs certainly do not depend on such concerns. Civilian customers and users, on the other hand, are often judged by bottom-line success and their futures are directly linked to such success. Put simply, defense markets focus on process and civilian markets focus on product. Risks taken or avoided reflect this difference.

Evaluate Hypotheses

Since this analysis is not premised on particular market sectors, the needs and beliefs hypothesized in the last two steps cannot be assessed empirically. However, my experience with my own company and many other companies supports these hypotheses. Moreover, some of the phenomena noted are frequently discussed in newspapers, magazines, and journals. Thus, to the extent these hypotheses are speculative—and I do not think they are—they reflect well-informed speculation.

Modify Alternatives

In the context of this example, modifying alternatives involves considering how, in general, a defense company should modify its offerings to compete in civilian markets. Based on the entries in the two NBP Templates, several differences between defense and civilian markets merit special attention:

- Maximizing performance vs. understanding performance needs,
- Cost as secondary vs. cost as primary,
- Satisfaction of requirements vs. impacting the bottom line,
- Innovation as a benefit vs. innovation as a risk,
- Process risks vs. product risks, and
- Context-specific evidence of previous successes.

The concern in this step of the analysis is determination of how a company's offerings to the marketplace should be modified in light of these differences.

A fundamental aspect of a company's strategy that must change is the way in which it approaches the market. The government advertises its needs in *Commerce Business Daily* and describes them in great detail in Requests for Proposals and Statements of Work. Civilian markets do not make their desires known in such straightforward ways.

To address civilian markets, companies have to research the market and gain an understanding of the benefits being sought. Methods for performing this type of research are described in Chapters 3–4 of *Design for Success* (Rouse, 1991). Based on such research, performance needs can be understood and appropriate technology used to enable provision of the desired levels of performance.

This market research should also address potential customers and users' perceptions of costs. This should not just include purchase prices, but also the costs of integration and support throughout a product's or system's life cycle. Also of interest is the extent to which potential customers and users will pay more for higher performance. While they may not be willing to pay the price of maximum performance, they may be interested in moving beyond baseline acceptable performance if the price is right.

The results of understanding the marketplace must be translated into products, systems, and services that balance performance and cost. Further, these offerings must provide indisputable benefits to customers and users, which usually means clear bottom-line benefits. Satisfaction of requirements is not sufficient if customers and users do not feel satisfied. Going beyond contractual requirements, without additional money, is often necessary to assure continued business.

The technology expertise that can provide enormous leverage in the defense market does not provide as much leverage in civilian markets. In-depth understanding of the market becomes more important than technology, at least from the perspective of marketing and sales. Technology may still provide a competitive advantage, but only as an enabling ingredient.

For the defense company in transition, this can be a difficult lesson. Many defense companies are heavily engineering oriented. Marketing is often a relatively small function, at least in comparison to most companies in civilian markets, especially consumer-products markets. In the process of transition, companies have to put increasing resources into marketing and relatively less into engineering. This can create substantial organizational stress, as discussed in Chapter 7.

One of the problems that this expanding marketing function has to deal with is the possible *negative* image of being technology innovators. This requires that innovation per se not be emphasized. Instead, the benefits of technology are emphasized, as well as the ways in which risks associated with these benefits are minimized. Chapters 7–8 of *Strategies for Innovation* (Rouse, 1992) illustrate a planning methodology that provides the requisite emphasis on benefits.

Risk is considered in quite different ways in defense and civilian markets. The Department of Defense, and the government in general, dictates and monitors the process whereby products and systems are designed, developed, and fielded. If contractors follow the prescribed processes, the customers assume most additional risks.

In contrast, civilian markets usually are not concerned with vendors' processes. They are not concerned with how a product or system is designed or developed, and they usually do not scrutinize companies' cost accounting. They simply want a low-price, high-quality product or system. The primary risks of concern, therefore, are associated with the product or system delivering benefits as promised.

For defense companies transitioning to civilian markets, this shift in perspective is important. Usually, defense contractors spend substantial amounts of overhead funds assuring compliance with government regulations about companies' processes. In civilian markets, these expenses result in a competitive disadvantage because, in effect, they are invested in managing risks that are not of importance to civilian customers.

A particularly difficult aspect of moving into civilian markets is the necessity of having context-specific evidence that the benefits sought by customers and users can be provided. As noted earlier, most customers and users do not set much store in evidence from other domains. Consequently, the first few sales in new markets can be extremely time consuming to obtain. The best people from marketing and sales are likely to be needed. In addition, very favorable prices, terms, and support agreements may have to be offered.

Modify Situations

The ways in which alternatives must be modified to facilitate transitions from defense markets to civilian markets are somewhat subtle and involve developing strengths that many defense contractors previously did not need. Much more subtle, however, are the ways in which situations should be modified.

One component of this involves educating the marketplace about the "new" company. If the company is large, with substantial name recognition, this requires a long-term program of marketing, promotion, and awareness-raising to change the market's image of the company. For small companies, the transition may, in effect, involve starting all over. This may sound easy, relative to what large companies may have to do, but can be just as difficult because many start-up skills and attitudes may have been lost.

Perhaps even more difficult is the process of educating the company about the "new" company. Well-learned self-images are exceedingly troublesome to change. Furthermore, those without an overall perspective of the company may have difficulty appreciating why the enterprise has to change course. Such phenomena are discussed at length in Chapter 7.

Plan Life Cycle

The transition from defense to civilian markets requires more planning to be successful. To a great extent, the plans of defense companies are dictated by the Department of Defense or the Ministry of Defense. Many defense companies' strategic plans, in my experience, are simply the list of defense procurements on which they intend to bid. In effect, many of these companies' marketing plans can be stated in terms of the Requests for Proposals whose preparation and release they are tracking.

Civilian markets typically require much more strategic thinking. Companies have to concern themselves with what benefits the market is seeking and likely to seek in the future, who is likely to attempt to provide these benefits, likely prices of offerings, and possible costs due to alternative process and product technologies. *Strategies for Innovation* (Rouse, 1992) deals with these and related issues in depth.

Summary

This section has considered understanding the marketplace as it relates to transitioning defense technology to civilian applications. Perhaps surprisingly, technology itself required little attention. The nature of the technology is, in most cases, much less of a problem than the differences between needs and beliefs in the defense and civilian markets.

This section illustrated how fundamental these underlying differences are. These differences dictate that companies trying to transition reconsider the ways in which they approach the marketplace. Civilian markets require understanding many subtle phenomena that either do not occur in defense markets or, due to the nature of government procurements, are made very explicit by customers and users.

While technology issues and differences are not central problems in transitions, organizational issues are. Not only is understanding the marketplace crucial—changing the organization to reflect the marketplace is also central. Consequently, we return in Chapter 7 to consider the problems associated with transitions from defense to civilian markets.

IMPLICATIONS OF ANALYSES

This chapter has considered in depth two applications of the NBP Analysis Methodology. In this section, lessons learned from these analyses are summarized. Figures 6.4 and 6.5 summarize the needs and beliefs identified in these analyses, beyond those gleaned from the compilations in earlier chapters. It seems reasonable to assert that many of these needs and beliefs are relevant to any situation that involves understanding the marketplace.

It is also useful to consider the results of these analyses in terms of other constructs introduced earlier, such as expectations, attributions, and mental models. For instance, people's expectations about the likely results of planning affect how they view planning tools. People might not value planning because they attribute past failures not to *their* lack of planning, but instead to uncontrollable external forces.

Expectations, attributions, and mental models also can be used to explain the great difficulty that defense contractors have in transitioning to civilian markets. Civilian markets do not react as they expect because their expectations are based on mental models developed in the course of operating in defense markets.

- Need to respond to the market
- Need to act quickly

- Need to perform better than competitors
- Need to achieve maximum performance

- Need to have own solution adopted
- Need to support technology

- Need to defend choices
- Need to avoid risks

- Need to recoup investments
- Need to make profits

FIGURE 6.4. Summary of needs identified.

- Belief in market-driven requirements
- Belief in value and quality
- Belief in performance

- Belief in off-the-shelf technology
- Belief in mature technology
- Belief in new technology
- Belief in inherent value of innovation

- Belief in "not invented here" (NIH)
- Belief in "not supported here" (NSH)

- Belief in reference sales
- Belief in domain-specific evidence
- Belief in sufficiency of meeting requirements
- Belief in formal validation

- Belief in short product life cycles
- Belief in long product life cycles

- Belief in the bottom line

FIGURE 6.5. Summary of beliefs identified.

Moreover, defense contractors tend to attribute this difficulty to external forces rather than their lack of understanding of civilian markets.

In both of these analyses, we can see a phenomenon discussed in Chapters 3–4. People tend to interpret their myths and models as facts. They fail to realize that their understanding of the roles and value of planning, or their understanding of how best to approach marketing, is inevitably biased and possibly flawed by the specific set of experiences they have had. Realization of this limitation is often all that is necessary to begin modifying expectations, attributions, and mental models.

SUMMARY

This chapter has focused on understanding the marketplace. Marketing and sales of software products were considered. Also studied was transitioning defense technology to civilian applications. The former analysis illustrated how to apply the NBP Analysis Methodology to the marketing and sales of a particular product. The latter analysis illustrated the problems associated with attempting a very substantial change of the markets that a company targets. Whether such changes involve moving from one sector of an economy to another, or moving from one country to another, important cultural differences can affect the success of the changes.

Beyond understanding the marketplace and differences among markets, a consideration of equal importance is understanding the culture of the enterprise making the changes. In-depth and accurate understanding of the market is a good start, but without specific consideration of how best to enable the enterprise to change, it is unlikely to be sufficient for success.

REFERENCES

Rouse, W. B. (1991). *Design for success: A human-centered approach to designing successful products and systems.* New York: Wiley.

Rouse, W. B. (1992). *Strategies for innovation: Creating successful products, systems and organizations.* New York: Wiley.

Chapter 7

Enabling the Enterprise

Understanding the market is essential, particularly with increased competition as virtually all countries attempt to succeed in global markets. The types of analyses described in Chapter 6, in conjunction with the product planning and business planning methodologies in *Design for Success* (Rouse, 1991) and *Strategies for Innovation* (Rouse, 1992) respectively, provide the means for a thorough understanding of the market. Unfortunately, while such understanding is necessary, it is not sufficient for success.

In order to transform knowledge of the market into business success, one needs an enterprise with appropriate aspirations, abilities, and resources. Often, an enterprise must change or adapt to form aspirations, gain abilities, and allocate resources so as to take advantage of market understanding. The process of changing and adaptation can be quite difficult, both in terms of recognizing the need to change and accomplishing the change.

This chapter is concerned with enabling the enterprise to recognize situations in which change is needed, identify the types of change needed, and implement changes. The discussions in this chapter, especially the discussions of the case studies, serve to illustrate how change affects an organization. These effects include individual and organizational reactions to forces for change and the need for organizational learning. By analyzing the needs, beliefs, and perceptions underlying these reactions, insights into how to facilitate change are gained.

FORCES FOR CHANGE

Why is change an issue? I think the driving force is the competition for resources and opportunities. Desirable resources are scarce. Many people want opportunities to raise their standards of living, and in particular the standards of living of their children.

I have seen these forces most clearly in developing countries. In remote villages in South Africa and Zimbabwe, dusty villages on the Bolivian Altiplano, and busy bazaars in Turkey and India, I have seen and talked with a wide variety of entrepreneurs who are determined to better themselves. They want change.

More specifically, they all want economic growth in which they can participate. Often such growth is facilitated by technological progress, particularly in societies that are open to change (Mokyr, 1990). Put simply, everyone is seeking prosperity, and those who embrace change as part of this process tend to be more likely to succeed.

In the developed countries, the forces for change are often different. Desired changes may be remedial in nature, with emphasis on fixing problems or tuning processes. Changes may also be adaptive in response to external changes or demands. These types of change can be viewed as aspects of organizational learning (Conlon, 1983), a topic discussed at length later in this chapter.

Developed countries are now also facing the need to change in response to crises. As Thurow (1992) discusses, economic crises have led to much discussion and debate about the nature of changes needed. While there are a variety of prescriptions for how we should respond to such crises, there is, at the very least, agreement about the fact of the crises.

In responding to the perceived economic crisis, there is a central tension between flexibly initiating change and assuring continuity with the past. If the enterprise opts for maximum flexibility, it is likely that people will perceive individual security as substantially diminished (Kanter, 1989). If the nature and importance of the crisis are emphasized to motivate change, there is the risk that people will start to view the world as a blank slate upon which virtually anything could happen (Wilkins, 1989). Clearly, there is a need to balance the value of the past, especially the psychological value, and the need to create a different future.

The resulting balancing act creates a dilemma. The past, especially for big companies, comes with much baggage (Kanter, 1989). The future may not be able to sustain this baggage. According to Tom Lloyd (1990), "If change is fast enough, big companies become the lowest cost producers of obsolete products." The difficulty is that these obsolete products may be the necessary result of baggage such as outmoded manufacturing facilities and employees with yesterday's skills.

The situation faced by big companies in developed countries is further complicated by the fact that necessary changes in the marketplace require changes of aspirations, aptitudes, and belief systems of managers (Lloyd, 1990). Such changes inevitably encounter barriers due to power, achievement, affiliation, and avoidance motives (McClelland, 1987). These types of barriers are discussed later in this chapter in terms of needs, beliefs, and perceptions.

How can enterprises be enabled to change and respond successfully to the crises and challenges that confront them? Many contemporary management gurus argue for the importance of a compelling shared vision of the enterprise's future, with special emphasis on the ways in which the future will be more positive than the past. It is difficult, and inappropriate, to refute this assertion. What is needed, however, is much more specific guidance for how to proceed.

ORGANIZATIONAL LEARNING

Successful organizational change inherently involves organizational learning. This section considers the nature of organizational learning, barriers to such learning, and means for overcoming these barriers. This material provides an important perspective with which to pursue the two case studies later in this chapter.

Chris Argyris and his frequent collaborator Donald Schon have studied professionals and their organizations for several decades. These studies have led to a simple, yet profound, theory of organizational learning. They state, "Organizational learning involves the detection and correction of error" (Argyris and Schon, 1978). In this process, two types of learning are possible.

Single-loop learning reflects corrections that permit the organization to carry on with its present policies to achieve its present objectives. Single-loop learning involves fixing or tuning the organization so as to improve its current ways of doing things. It does not involve rethinking the nature of the organization.

Double-loop learning reflects corrections that involve modification of the organization's underlying norms, policies, and objectives. Rather than fixing or tuning, double-loop learning involves remodeling or replacing various aspects of the organization. Goals, strategies, and plans are reconsidered and fundamental changes are entertained.

Argyris and Schon argue that organizational learning takes place in the context of "theories of action." They assert that "all deliberate actions have a cognitive basis that reflects norms, strategies, and assumptions or models of the world." Thus, we again see the construct of mental models invoked to explain behavior.

They contrast espoused theories vs. theories-in-use, and quite reasonably claim that theories-in-use are what enhances or inhibits learning. If the theories-in-use in an organization tend to be totally goal oriented, focusing on wins and losses and not allowing explicit discussion of negative perceptions, it is unlikely that double-loop learning can occur. In contrast, if theories-in-use emphasize collecting valid information, making free and informed choices, and fostering internal commitments to choices, double-loop learning is much more likely.

Using our earlier terminology, the mental models of members of an organization determine its potential for learning. Rational, goal-oriented, "can do" models are appropriate to the extent that single-loop learning is sufficient for organizational success. However, when double-loop learning is needed—when fundamental changes are necessary—such models can be barriers to learning.

Barriers to Learning

There are a variety of barriers to organizational learning. Argyris and Schon (1978) suggest that withholding negative information, avoiding direct confrontation, and seeking unilaterally to control situations tend to inhibit learning. Information about what has gone wrong and negative perceptions are often what drives learning. In their absence, diagnostic possibilities are severely limited. If, in addition, those aware of the situation do not involve others in the resolution, the learning experience is limited to, at most, a few people.

Argyris (1982) discusses two other inhibitions to learning. One is *distancing*, whereby people do not accept responsibility for problems, and also do not accept responsibility for solutions. The second inhibition is *disconnectedness*, which occurs when people are not aware of the reasoning underlying their actions. In other words, they are unaware of their theories-in-use and the relationships between these theories and their actions.

Peter Senge (1990) discusses what he calls organizational learning disabilities, based in part on the work of Argyris. One type of disability is the tendency of people to see themselves in terms of their position rather than their aptitudes, abilities, and skills. As a result, organizational changes that require new positions, and the loss of old positions, can cause people to lose their self-image.

There is also a tendency to blame what Senge calls "the enemy out there." This relates to the fundamental attribution error discussed in Chapter 2. If somebody else is always to blame, there is little that you can do to avoid repeated problems. While we can "take charge" to deal with these problems, we cannot possibly achieve double-loop learning.

There also tends to be a fixation on events, often problem events. However, many problems and undesirable trends manifest themselves as slow, long-term processes that are difficult to deal with once they result in "events." Senge discusses such processes in terms of the parable of the boiled frog. It is not likely that a frog can be dropped in boiling water without it attempting to jump out. However, if the water is heated very slowly, the frog will not notice the slow increase of water temperature until it is too groggy to climb out.

Many organizations find themselves in similar situations. Overhead budgets slowly grow, a larger and larger portion of the budget is nondiscretionary, less and less attention is paid to existing and potential customers, and suddenly there is a crisis. Attempts are made to cut overheads, as well as redirect resources. However, the people and "turf" involved resist, and the frog slowly gets boiled.

Senge also argues that the notion of learning from experience is often a delusion. The consequences of important decisions are typically many years displaced from the decisions. Without processes that carefully consider consequences and, equally important, measure outcomes and relate them to previous decisions, it is difficult to detect the errors that drive the learning process.

Finally, Senge discusses the myth of the management team. Citing the work of Argyris, he notes that most management teams are not open to dealing with negative information and the possibilities of conflict. As a result, the team deals well with routine problems, but may become dysfunctional in crises.

Senge argues that these disabilities are particularly problematic in diverse, cross-functional teams. In such situations, each member has his or her own linear mental model. Each model has a different focus, such as, marketing vs. manufacturing. Each model emphasizes different causes and effects. Hence, it can be very difficult to construct a shared picture.

I have often encountered this problem in my seminars and workshops on product planning and business planning. I usually coach the group of participants through a few exercises that typically lead to their mutual discovery of things about their enterprise that they had not realized. These discoveries often prove central to the group's subsequent abilities to develop new plans for market and organizational innovations.

Alan Wilkins (1989) also considers organizational characteristics that constrain learning. Problematic characteristics include assuming that past competence is a guaranteed formula for success, thinking in stereotypes and labels, getting into ideological ruts, and developing superstitious commitments to certain practices and actions. Obviously, these characteristics imply theories-in-use that will strongly inhibit double-loop learning.

Wilkins also discusses problems that inhibit change despite the recognition that it is necessary. He notes that resistance to change is often due to uncertainty about how to change rather than denial that change is needed. Employees can feel powerless in the process of change. Lack of motivational faith—see Figure 3.11—can hinder employees' commitment to change.

Overcoming Barriers

Edward Conlon (1983) discusses a process model for implementing change, based in part on the work of Kurt Lewin. The first step of this process involves determining the beginning and end states. This requires that the organization know both where it is and where it wants to be. As the case studies later in this chapter illustrate, assessing the current state and envisioning the desired future state can be quite difficult.

The next step in this process concerns unfreezing old behaviors—for instance, unfreezing an organization's mental model of the marketplace. This involves considering individual, social, and organizational influences. This step is much easier if the changes envisioned are perceived to be rewarding or potentially rewarding.

The third step in Conlon's process concerns moving behaviors—for instance, creating a new mental model of the marketplace. Training is often an important element of facilitating such changes. Training can provide a means of practicing the desired behaviors more frequently than may be possible in the normal course of events.

The next step involves refreezing behaviors. This can be accomplished by explicit and consistent reinforcement of the desired behaviors. Organizational leaders can also create opportunities to review and reflect upon the importance of the change.

The last step in the change process concerns evaluating the end state achieved. This involves asking whether the results are as good in fact as they were in fancy. There may be, for example, unanticipated side effects or unexpected costs. Further changes, and uses of the change process, may be needed.

In my experience, the most difficult part of this change process is gaining consensus on the current state and the desired future state. Considerable exploration may be needed to accomplish this step. In such pursuits, Senge (1990) suggests that members of management act like researchers and designers, rather than traditional, hierarchical bosses.

John Brown (1991) asserts that research can provide the means for reinventing the corporation. He argues that organizational learning can be facilitated by studying the "communities-of-practice" within the enterprise (Brown and Duguid, 1991). By studying itself, Brown concludes, an organization can uncover much latent innovation.

Senge (1990) suggests that a primary benefit of planning is the learning that results. In my seminars and workshops on product planning and business planning, I usually find that participants discover from each other much about their current state and alternative future states. Participants often tell me that the dialogues that occur in these sessions seldom, if ever, occur as part of their normal work processes.

Summary

This section has provided a brief overview of forces influencing change in enterprises and the nature of organizational learning that must underlie adaptation to these forces. The remainder of this chapter considers two case studies using the Needs–Beliefs–Perceptions (NBP) Model, Template, and Analysis Methodology, as well as the concepts and principles discussed in this section.

The first case study focuses on the organizational difficulties associated with changing from service to product markets. As with the case study of marketing and sales of software products in Chapter 6, this case study is drawn from my own company. This study, therefore, focuses on how a small company dealt with major organizational changes.

The second case study is concerned with changes from defense to nondefense markets. Drawing on my many experiences of helping small, medium, and large companies consider this type of change—see Figure 1.4 for a compilation of the types of companies—an NBP analysis is used to illustrate the substantial difficulties associated with this change.

CHANGING FROM SERVICE TO PRODUCT MARKETS

In Chapter 6, the transition of Search Technology from selling primarily "one off" software systems to selling off-the-shelf software products was discussed. An NBP analysis was used to understand how to market and sell these products. This analysis focused on the perceptions of customers and users.

In this section, the same transition is considered from the perspective of the management and employees of Search Technology. Enabling this enterprise to make this transition proved, in many ways, to be more difficult than understanding new markets and appropriate approaches to those markets. In fact, the process of changing Search Technology continues at the time of this writing. Fundamental change can take a long time.

To discuss this transition from the perspective of management and employees requires a little more background. Many of the software systems developed by the company were not sold as systems per se. Instead, engineering services were sold in terms of the person-hours necessary to develop the system of interest. Often the system delivered was a "proof-of-concept" prototype.

Consequently, customers were typically buying person-hours, not systems, and the system that resulted mainly served to prove that the technology of interest could provide the benefits sought. While such efforts were sometimes risky in terms of the technology breakthroughs attempted, these risks were not borne by the company. The company's primary risk was lack of follow-on business with customers if they were not satisfied. For government customers, this risk was often minimal, since relationships and goodwill are not allowed to play as large a role in government procurements as they do in nongovernment markets.

This sounds like a "sweetheart deal"—opportunities to work with leading-edge technologies, without having to deliver the finished systems and few risks if expectations are not met. However, it can also be extremely frustrating. People know that what they create will only be demonstrated, not actually used. Excellent work often does not lead to future opportunities with customers and users. Projects get delayed, postponed, started, stopped, canceled, and occasionally resurrected.

Not surprisingly, people often talked about focusing on products rather than services. If we had successful software products in the market, we would be able to see the impact of our work more directly and tangibly. People see Apple

Computer and Microsoft and say, "Let's do what they did!" I am always amazed at the number of software people who are sure they could be the next Steve Jobs or Bill Gates—they just haven't got around to it yet.

From this perspective, you might think that Search Technology's developing, marketing, and selling of its family of Advisor products would cause great enthusiasm among managers and employees. It did not. In fact, while managers were reasonably supportive, many employees were neutral at best. Only a few of the software staff were at all enthusiastic.

I wondered about the basis of this response, which was very surprising to me. There were many reasons for it. First, and perhaps foremost, the Advisors did not require the latest and greatest technology. High-end microprocessor-based platforms were targeted, rather than state-of-the-art engineering workstations, because that was the computer of choice of the customers and users of these products. Similarly, the software environment chosen was new, but not as sophisticated as our software staff would have liked. All in all, many people eschewed the Advisors as "low tech."

Another problem was the fact that our products would have to be supported. User's manuals were needed. A product support function to answer telephone calls was established. Software testing and usability testing became important elements of the company's technical activities. Many members of the software staff did not want to have anything to do with these types of support functions.

At the same time, the company's work in software systems also changed. Instead of providing high-tech, proof-of-concept prototypes, installed systems were now the typical deliverables. In shifting from government markets to non-government markets, the company found that customers and users were only interested in proofs of concept as steps on the path to completed, installed systems. Thus, the emphasis on completed and supported systems increased in the services segment of the business also.

This emphasis, for both products and services, led to much organizational stress. In the context of the discussions in Chapter 3, many employees perceived that the company had changed organizational denominations. For some employees, I am sure there was what Rosabeth Moss Kanter terms a "crisis of commitment" (Kanter, 1989). The company seemed to be changing in fundamental ways.

Before proceeding with an NBP analysis of this problem, it is useful to clarify the essence of the problem being addressed in this analysis. The problem is one of designing, or actually redesigning, a company. This redesign was prompted by a need to move away from shrinking government markets and a desire to sell products in the civilian marketplace. This redesign was problematic in that many employees wanted the benefits of changing without having to make any fundamental changes. As such changes began to emerge, conflicting perceptions of the viability, acceptability, and validity of these organizational trends also emerged. The analysis in the remainder of this section is focused on understanding the basis of these differing perceptions.

Identify Measures

Viability, acceptability, and validity mean very different things from the point of view of top management vs. the perspective of employees. Management necessarily focuses on company-wide benefits in terms of presence in the market, sales, cash flow, and profits. Costs include those associated with development, marketing, sales, and support of products and systems, as well as fixed costs for facilities, equipment, salaries, and benefits.

From employees' perspectives, benefits include salary, health insurance, and so on. Of perhaps greater importance, however, are the working conditions and the nature of the work. I have found that many engineers and scientists take salary, insurance, and other benefits for granted. They attach greater value to being able to work with leading-edge technology without any constraints.

With regard to perceived costs, employees are concerned with the difficulties of discarding old skills and gaining new ones. Engineers and scientists also often associate costs with having to deal with the organization and any constraints that the organization imposes. Change implies costs in general, compared to maintaining current activities.

Concerning acceptability, top management, if it is doing its job, should be an agent of change. Acceptable changes are those that are compatible with the enterprise's strategic plan and do not entail excessive risk. Acceptability usually also requires that changes not detract from the enterprise's image, both internally and externally.

For employees, perceptions of acceptability relate to the likely impact changes will have on their roles. If changed roles are perceived to be consistent with their self-images, or the images to which they aspire, acceptability problems are unlikely. Employees are also often concerned that changes are socially acceptable to their peers, both within and outside the company.

Valid changes from management's point of view are those that will work in the sense of solving the problems for which they were designed. Typical concerns are whether planned changes will increase sales, lower costs, solve quality problems, or ease constraints. As noted in Chapter 6, validity is often judged on the basis of experiences of other enterprises with the same types of change.

As with the other issues, employees' concerns usually center on their jobs. Changes are perceived to be valid by employees if they result in secure, high-paying jobs. If planned changes result in much uncertainty about job security, then perceptions of validity will be low, or at least withheld until uncertainty is lessened. These perceptions are usually quite personal. People are not completely assured by knowing there will be jobs in general—they want to know about their jobs in particular.

Determine Alternatives

The alternative of interest in this case study is a specific type of company. This company primarily sells off-the-shelf products and installed systems. These offerings drive all of its subordinate goals in manufacturing, services, and R&D. Chapter 7 of *Strategies for Innovation* (Rouse, 1992) describes the structure and functioning of such a company in detail.

The company that this new company replaced—the old Search Technology—focused on technical expertise and excellence as its primary value added. R&D was supposed to produce technology that would somehow, but never did, yield new products. The old company believed in "technology push," while the new company is committed to "market pull."

Assess Perceptions

As noted earlier, measures are identified by first considering viability, then acceptability, and then validity. However, perceptions are assessed in the opposite order. Therefore, validity is the first concern.

From the perspective of top management, the change from service to product markets was inherently valid, acceptable, and viable because it provided an avenue that better served the company's mission. The change had emerged from plans initiated several years before and updated with each successive year. Progress had been slow but steady. All employees had been informed, and many had been involved, with these plans. While the implications of these plans had often been discussed, for most people they did not seem real until fundamental changes became manifest.

As noted earlier, employees' reactions were mixed. Most perceived the changes to be valid from a company point of view, and necessary due to market trends. The administrative staff perceived the emerging changes to be acceptable and it appeared to trust in their viability. Since this staff was cut by 50 percent in the process of change, its remaining members had been assured that their redesigned jobs—redesigned by them—were clearly part of the company's future.

Many technical staffers had negative perceptions of the acceptability of the changes. As discussed earlier, they did not want to be involved with other than leading-edge technology and wanted no part of product support. For several individuals, it appeared that their self-images as creators, or at least users, of state-of-the-art computer software and hardware would not be supported by being involved with development of software products for microcomputer applications in businesses rather than, for example, in airplanes.

For these people, viability was also questionable. While they appeared to feel that the benefits might outweigh the costs for the company as a whole, they did not see much benefit for themselves and perceived sizable costs associated with having to produce completed systems and products. The fact that the results of their efforts would be providing benefits to customers and users was certainly positive, but not enough to balance the potential changes in their jobs and roles.

Define NBP Templates

This analysis potentially could involve one NBP Template for each type of stakeholder, that is, management, administrative staff, technical staff, bankers, and shareholders, all of whom have a stake in the future of the company. For the purposes of this case study, however, only one template is needed. This template (Fig. 7.1) is used to summarize the positive and negative perceptions of the technical staff regarding the change from service to product markets. Using this template, the focus of the remainder of this analysis is on understanding and improving these perceptions.

Hypothesize Beliefs

The beliefs indicated in Figure 7.1 were compiled using two sources. First, the understanding of the problem elaborated in earlier steps of the analysis was reviewed. Second, the entries in Figures 3.6, 3.11, and 4.3, particularly the latter two figures were considered. Where appropriate, the wording of beliefs was changed to fit the context of the problem.

The beliefs associated with positive perceptions reflect a sense that the organization can succeed with its product plans. While this will result in a different and perhaps new enterprise with new roles to be filled by the technical staff, this is viewed as a desirable opportunity. The benefits possible with these changes are seen as outweighing the costs of changing and dealing with uncertainty and new demands.

Negative perceptions are associated with a different set of beliefs. The value of the products being introduced is questioned, as is the possibility of understanding the marketplace. Necessary new skills are seen as undesirable, or at least difficult to acquire. The cost of dealing with the likely changes is thought to be much too large.

Hypothesize Needs

Using Figures 3.5, 3.7, 3.10, and 4.2, particularly the latter two figures, needs underlying the beliefs in Figure 7.1 can be inferred. Also important in this in-

ATTRIBUTE	PERCEPTIONS	BELIEFS	NEEDS
Viability	Positive	Opportunities, salary, benefits, etc., will outweigh costs of change, uncertainty, and new demands	Need to grow, create, and have impact, need for salary, etc., lack of need to avoid change and uncertainty
	Negative	Cost of change very large	Need to avoid change, uncertainty, and possible failure
Acceptability	Positive	New enterprise and new roles desirable and achievable	Need to belong and achieve, need for self-actualization, need to maintain self-image
	Negative	New skills required and difficult to acquire	Need to exercise skills, employ particular methods, and belong to discipline, need to avoid possible failure, need for continuity
Validity	Positive	Plan will work, organization will prosper, and roles are clear and meaningful	Need for affiliation, job security, and control
	Negative	Skeptical about products, marketplace inscrutable	Need to avoid uncertainty and possible failure

FIGURE 7.1. NBP Template for changes from service to product markets.

ference process is an understanding of the context of the problem as outlined earlier in this section.

Comparing the needs associated with positive perceptions and those associated with negative perceptions, it appears to me that the primary difference is the relative emphases on particular types of needs. On the positive side, needs to grow, create, achieve, and be part of a new venture appear to dominate and result, in effect, in a lack of needs to avoid change and uncertainty. On the negative side, avoidance needs are dominant, as are needs to exercise hard-earned skills and use familiar methods.

My hypothesis, at the time, was that all of the needs indicated in Figure 7.1 were, to some extent, relevant to all of the people involved. The difference between the two groups was one of emphasis. Those who emphasized the "upside" of the changes perceived on balance that the changes were good. Those who focused on the "downside" perceived the changes as bad. The challenge, therefore, became one of portraying changes in a manner that enabled as many people as possible to see and value the "upside" for them.

Evaluate Hypotheses

The hypotheses summarized in Figure 7.1 were evaluated over several months. Much of this evaluation was informal, involving many discussions with numerous employees. While these conversations were useful, they did not result in definitive testing of the hypotheses about needs and beliefs. Nevertheless, many issues were raised.

These issues, such as, the role of R&D in the "new" company, provided the basis for a series of brief (three-to-five-page) "thought" papers that were circulated, with a request for feedback. Many questions and comments resulted. This led to a variety of focused conversations that, in effect, dealt with the types of needs and beliefs noted in Figure 7.1.

This process could not, by any means, be justified as a rigorous test of the hypotheses. No scientific paper could be published based on such informal methods. However, the evidence was more than adequate for making a business decision concerning what to do about negative perceptions.

Modify Alternatives

This step concerns modifying alternatives—in this case, the "image" of the new company—so as to satisfy needs and not conflict with beliefs. This was accomplished, albeit imperfectly, by a combination of three tactics. First, we emphasized continuity with the past. While fundamental changes were intentionally being made, links to past strengths and accomplishments, it was argued, provided the foundation that enabled these important changes.

Second, we focused on clarifying new and evolving roles. This included showing how our "traditional" skills fit into the future and what new skills would be needed. We also committed the company to providing any training necessary for gaining these new skills. In fact, little training was needed in the short term. What was needed was the commitment to provide training when it was necessary.

The third, and most important, tactic emerged from the focused discussions noted earlier. We had made a concerted effort to articulate the "vision" of the new company in terms of targeted markets; benefits provided to customers and users in these markets; products, systems, and services that would provide these benefits; and projected sales and profits. Many members of the technical staff did not find this vision compelling.

After much discussion of this problem, it struck us that we had been emphasizing an "outward" vision that focused on why the marketplace would be committed to the company. To keep employees committed to the company, we needed an "inward" vision. People wanted to know what everyday life at Search Technology was going to be like. This included the size and composition of the staff; the technologies, methods, and tools that would be employed; the nature of the facilities and work environment; and probable salaries and benefits.

My natural predilection is to be almost totally customer oriented—the inward vision will be "whatever it takes" to fulfill the outward vision. Consequently, at first I had difficulty characterizing the inward vision. To overcome this difficulty, I resorted to a simple mechanism.

I wrote a quite brief and simple story about what day-to-day life was like five years in the future. I then circulated this three-page story to a half dozen employees representative of all levels of the company. I asked them to read this story and write down any questions that the story prompted. The result was many, many questions. In two or three cases, the list of questions was longer than the story.

I then revised the story so that all the questions were answered, often indirectly, by the characters and events in the story. The expanded story was then circulated to the whole company, again with a request for questions. The new round of questions led to further elaborations of the story. In effect, the process was one of finding out what type of story was meaningful to the company's employees. While we were not open to employees totally dictating the contents of the story, we were very open to their telling us what type of story was meaningful and desirable to them. After several iterations, the story had grown to six to seven pages and virtually everyone in the company had influenced its content.

Modify Situations

By modifying and clarifying the characterization of the "new" company, it was possible to change or moderate many negative perceptions. However, it was not at all possible to eliminate concerns about the difficulty of change and the inherent uncertainty. The company was in the process of writing a new story that was not simply a rewrite of the old story.

For the longer term, modifying the situation involves modifying the company's story. More specifically, the long-term tactic was to get people, slowly but surely, to realize that the new company was a natural outgrowth of its 12-year path up to that point. The goal was to create a climate that would foster celebration of the important changes that the company was planning and experiencing.

Plan Life Cycle

Enabling the enterprise is likely to involve constant change. I used to think that we would someday "arrive." Now I know that not only will we not arrive—we should avoid such a possibility.

It is easy for the culture of the enterprise to become static. It then becomes very difficult to gain any momentum for change. The larger the enterprise, the greater the difficulty. As Thurow (1992) has pointed out, initiating change in such situations may require a crisis.

What is needed is an enterprise that can continually and creatively change in order to overcome obstacles and take advantage of opportunities. This climate of continual change should be created in the context of an enduring set of needs and beliefs. By maintaining and nurturing a core set of values, change can be built upon a stable foundation.

Unfortunately, it is easy to lose track of core values and focus on the surface features of changes. As noted in Chapter 11 of *Strategies for Innovation* (Rouse, 1992), one of senior management's most important roles involves communicating and reinforcing the enterprise's values. From personal experience, I know how difficult and endless this task can be. Nevertheless, it is absolutely central to being able to sustain creative change.

Thus, planning the life cycle of change is a process of initiating and completing an endless series of changes. Each change, hopefully, reflects something new that the enterprise has learned. Each change also establishes the preconditions for further learning and subsequent changes.

Summary

In this section, the case study described involved a small software company changing from service to products markets. This would seem to be a straightforward change, requiring the same technologies and many of the same core competencies. However, this change required much more effort than expected and much more time—roughly three years.

While this experience may have been unique, I doubt it. My work with a wide range of companies suggests that change is surprisingly difficult. In the next section, one of the most difficult, and most important, types of corporate change is discussed.

CHANGING FROM DEFENSE TO NONDEFENSE MARKETS

In Chapter 6, differences between defense and civilian markets were discussed in terms of differences in needs, beliefs, and perceptions of customers and users in these two types of markets. Understanding these differences is critical to being able to offer civilian markets the kinds of products, systems, and services that will succeed. However, as indicated in the discussion of the importance of this type of understanding, such understanding is only a starting point for change. An equally challenging, or perhaps more challenging, requirement is creation of the organizational changes necessary to succeed in nondefense markets.

Enabling defense enterprises to change from defense to nondefense markets is a critical issue in virtually all countries. Economic prosperity depends on transitioning financial and intellectual capital from building weapon systems to creating products, systems, and services that delight the global marketplace. Among the remaining economic superpowers, the United States faces perhaps the biggest challenge because our economy has depended on defense expenditures to a much greater extent than Japan and Germany (Thurow, 1992).

Successful transitions from defense to nondefense markets depend on understanding the essence of the problem. For the most part, the problem is not a question of technology. Many defense technologies and most core technical competencies are relevant to civilian markets.

To an extent, the problem is one of understanding the marketplace, as discussed in Chapter 6. However, my experience is that this problem can be readily overcome. It is quite clear what must be done to understand new markets and determine how to delight these markets.

The most daunting problem, I think, is changing the nature of the enterprise so as to be able to make effective use of the newfound market knowledge. Using the terminology introduced in Chapter 3, the requisite changes involve changing organizational denominations, rather than just "tuning" the existing denominational structure and culture. The consequences of changing organizational religions can result in what Kanter (1989) calls a "crisis of commitment." This section explores the nature of this crisis using the NBP Model, Template, and Analysis Methodology.

It is useful to differentiate between the types of organizational change necessary to become better military units and military contractors, and the change required to change markets. Bob Blanchard and Bill Blackwood (1990) consider the former problem using a version of Conlon's (1983) model, which was discussed earlier in this chapter. Their work is part of a major military effort headed by Harold Booher (1990) called MANPRINT whose goals include enhancing the quality of weapon systems. This chapter is not concerned with this type of change.

Another type of organizational change involves transitioning from defense markets to other government markets. As noted in Chapter 6, this is usually an easier change than transitioning to private-sector markets. However, this type of change can only work for a minority of defense companies. Quite simply, the government cannot be the primary source of job creation if we are to succeed in the increasingly competitive global marketplace. This type of change is not considered further in this chapter.

The organizational changes of interest, therefore, involve defense companies changing from defense markets to private-sector markets. The necessary changes are substantial. As noted in Chapter 6, one change concerns a different approach to marketing and sales which requires substantial investment in this function.

To be specific, my experience is that the cost of sales for defense markets averages 1–2 percent of the total value of the sale. The cost of sales includes the expenses of some market research, sales materials and presentations, and bids and proposals. In contrast, private-sector markets often have costs of sales of 20–30 percent of the total value of the sale. In this case, these costs include extensive market research, much advertising and promotion, and sizable sales commissions.

Another contrast concerns the costs of goods sold. In defense markets, especially domestic markets, the costs of designing, developing, and manufacturing typically include about 85–95 percent of the total value of the sale. In effect, defense contractors sell the government person-hours rather than products and systems. Consequently, the more labor-hours the better.

Obviously, private-sector companies could not survive if costs of marketing and sales were 20–30 percent and costs of goods sold were 85–95 percent. Hence, the direct labor content of products and systems has to be much less, perhaps 20–30 percent. One of our customer's products requires only 1 percent for the costs of goods sold—in other words, their cost of raw materials and manufacturing is one percent of the selling price. Their costs of sales are about 70–80 percent.

As a consequence of these differences, engineering functions are, by no means, as dominant in nondefense companies. While engineering and technology-related functions can still be central to the enterprise's success, technology and technical expertise no longer are alone on center stage. Not surprisingly, this change of status can result in considerable organizational stress, particularly for "high tech" companies.

Another contrast concerns administrative and other overhead functions. Working for the government usually requires extensive cost accounting and contract administration functions that can be much leaner in nongovernment markets. The government monitors costs closely because the prices that contractors are allowed to charge are based on their costs. The goal is to make sure that all costs are justifiable. Any costs that are not found to be allowable are deducted from the amounts paid to contractors by the government.

Contract administration is also complicated by the fact that contract clauses sometimes reflect social agendas that go beyond the essence of the work being done (Rouse and Boff, 1987). For instance, we have often been required to certify—by filling in and signing a form—that the software products we deliver to the United States government do not include any non-United States made jeweled bearings! While this example is extreme, it serves to illustrate the extra paperwork that defense contractors must create organizational functions to handle.

In a special issue of *IEEE Spectrum* (1992) dealing with the defense industry, it was estimated that requirements such as just outlined raise prices by 15 percent. This could amount to 20–30 percent of all overhead. If defense companies maintain these functions as they attempt to succeed in private-sector markets, they face a competitive disadvantage. The obvious answer is to reduce staffing dramatically in the areas of cost accounting and contract administration. As might be expected, such staffing reductions are another contributor to organizational stress.

There is another, much more subtle, stressor associated with changing from defense to nondefense markets. Quite often, people in defense companies have developed self-images of working on the cutting edge of high technology. They see nondefense as low tech.

In the past few weeks, I have encountered this claim in three companies involved with military transport aircraft. In response to this assertion, I have used Lester Thurow's (1992) observation, "The technology of subsonic aircraft is forty years old. World-class reservation systems are on the cutting edge of technology." People find this observation very difficult to accept. In one case, the people involved have been building the same aircraft for almost 40 years. Nevertheless, they see themselves and their company as high-tech innovators.

We could try to convince them that they are wrong, but this tactic does not lead to positive results. While we might think they have misperceptions, they feel that they simply have different perceptions. And, their perceptions can cause them to feel an impending loss of status as their company tries to change to nondefense markets.

The discussion thus far in this section has served to show that there are many sources of negative perceptions that can undermine attempts to make major organizational changes. In the remainder of this section, these perceptions are considered using the NBP Model, Template, and Analysis Methodology. While this analysis does not lead to any "magic" solutions, it does provide a few insights that can potentially contribute to dealing with this crucial problem.

Identify Measures

Chapter 6 considered the problem of transitioning defense technologies from the perspectives of customers and users. In this chapter, we are concerned with the points of view of employees and management, as well as, to a slight extent, shareholders. This step of the analysis involves identifying measures of viability, acceptability, and validity for these three types of stakeholders.

Considering viability, benefits include sales, cash flow, and profits, which translate into payroll, fringe benefits, dividends, and capital gains. Beyond the usual expenses of operating an enterprise, costs include those associated with making changes. These costs are partially monetary, and partially social and psychological.

Measures of acceptability relate to the extent to which new directions are compatible with the existing culture and self-images. Substantial change may not be acceptable, particularly if the likely results of the change are undesirable in terms of status, power relationships, and self-images. Acceptability also decreases when people perceive increased possibilities that they might not fit into the enterprise's plans.

Validity measures relate to perceptions of the potential of the transition plan. If people perceive that the plan is well conceived and that the enterprise is capable of successfully executing the plan, they are likely to perceive validity to be high. On the other hand, if the enterprise's current difficulties—which are likely to be a primary motivation for change—have caused people to question the organization's abilities, people may be skeptical about the chances of successfully implementing plans.

Determine Alternatives

This step involves defining the alternatives about which perceptions of viability, acceptability, and validity are to be assessed. For this case study, the "alternative" is the new enterprise, pursuing its business in nondefense markets. Several attributes can influence the nature of the new enterprise.

Figure 7.2 indicates two of these attributes. The "nature of innovation" differentiates enterprises whose product is innovative from enterprises whose innovative processes are used to manufacture products designed by others. The "role of company" attribute contrasts enterprises whose products or processes contribute to some other enterprise's end products with enterprises who sell end products themselves.

Another attribute is the maturity of the technology involved. Mature technologies are those that have repeatedly proven themselves, where most of the lessons to be learned have been learned. Immature technology may be promising, but has yet to prove itself.

Two additional attributes include the size of the enterprise and its financial trends. Bigger enterprises have more resources, but they also have more inertia and can have bigger problems. Positive financial trends are usually easier to build from, but negative financial trends are often associated with enterprises seeking to change.

Role \ Innovation		Nature of Innovation: Product	Nature of Innovation: Process
Role of Company	Supplier	• Design constrained to be compatible • Emphasis on propietary technology and mfg quality	• Design dictated by customer • Emphasis on mfg cost and quality
	End Product	• Design driven by end users and customers • Emphasis on design, mfg, mkt, sales, services cost and quality	• Design dictated by mkt standard • Emphasis on mfg, sales, service cost and quality

FIGURE 7.2. Roles of companies and nature of innovation.

Changing from defense markets to nondefense markets involves not only identifying and understanding new markets, but also determining how the enterprise must be changed. The extent of this latter type of change should strongly influence which nondefense markets are chosen. For example, a process-oriented supplier of components based on mature technologies probably should not entertain thoughts of becoming a product-oriented seller of end products that incorporate state-of-the-art technologies. This is especially true if the enterprise is very large with strongly negative financial trends.

In general, three attributes indicate the degree of organizational change that is likely to be needed: nature of innovation, role of company, and maturity of technology. Substantial changes along all three of these attributes imply substantial organizational changes. The other two attributes—enterprise size and financial trends—are indicative of the amount of inertia to overcome and the resources available to do it. Small enterprises with positive financial trends are most likely to be able to support change, large enterprises with negative financial trends, least likely. Unfortunately, the former may not need to change, while the latter may have no choice but to change.

For the purpose of this case study, it is not necessary to specify the particular changes that an enterprise entertains. The analysis of the basis of positive and negative perceptions can be pursued using the material discussed thus far in this section. However, if we were considering the problem of changing markets for any particular enterprise, the analysis would necessarily have to consider specific organizational changes and perceptions of these changes.

Assess Perceptions

The first concern lies with perceptions of validity. To the extent that people perceive the enterprise's plans as highly likely to preserve their jobs and the enterprise's future, they usually perceive changing from defense to nondefense markets as a valid solution to the problem of declining defense markets. While there are always some skeptics, I have found that most people think, perhaps overoptimistically, that the enterprise can make this change successfully if it chooses to do so.

Perceptions of acceptability often constitute a major problem. Many people do not like being forced into new, uncertain roles. Typically, their self-images are entangled in their old roles. Moreover, they often are not confident that they can succeed in the new roles. Therefore, they may see changing markets as a valid solution to their problems, but not a desirable solution.

Viability is perceived in terms of benefits and costs. Benefits include sales, cash flow, profits, salaries and wages, fringe benefits, and so on. For most people, costs are related to the difficulty of changing and learning new skills. To the extent that benefits will not decrease and the costs of change are bearable, people will usually have positive perceptions of viability.

Define NBP Templates

This analysis employs one NBP Template (Fig. 7.3) which summarizes positive and negative perceptions regarding change. These perceptions are not associated with any single class of stakeholder (e.g., employees, management, or shareholders). They represent the integration of all these stakeholders' perceptions.

Hypothesize Beliefs

The beliefs column of Figure 7.3 was filled in by first reviewing the results of the earlier steps of the analysis and then considering the entries in Figures 3.6, 3.11, 4.3, and 6.5, particularly the latter three figures. The wording of beliefs was modified as necessary to conform to the context of this discussion.

Positive perceptions are associated with beliefs that plans will work, by drawing on organizational strengths and team support. Further beliefs include expectations that roles will be affected positively and benefits of changing will exceed costs.

Negative perceptions are associated with a much different set of beliefs. There is a lack of faith in the organization and its members. There are also beliefs that

ATTRIBUTE	PERCEPTIONS	BELIEFS	NEEDS
Viability	Positive	Benefits to company in terms of sales, cash flow, and profits will exceed costs of change	Need for cash flow and profit, need for company to prosper
	Negative	Cost of change very large, very difficult to change large organizations	Need to avoid change, uncertainty, and possible failure
Acceptability	Positive	Organizational strengths remain valuable, roles affected positively	Need to achieve, need for power, need to maintain self-image
	Negative	Power structure changed, new skills required and difficult to acquire	Need for power, need to avoid possible failure
Validity	Positive	Plan will work with organizational changes that are possible, team will support change	Need for power and affilitation
	Negative	Change not possible, marketplace inscrutable	Need to avoid uncertainty and possible failure

FIGURE 7.3. NBP Template for changes from defense to nondefense markets.

position and power will be lost and new skills will be extremely difficult to acquire.

Hypothesize Needs

The needs column of Figure 7.3 was filled in by consulting the usual sources—previous steps of the analysis and compilations of needs from earlier chapters. The entries in this column provide a partial explanation for the beliefs just summarized.

Needs for power, achievement, and affiliation, as well as very practical needs for sales, profits, and so forth, underlie the beliefs associated with positive perceptions. People who see the change from defense markets to nondefense markets as a means for accomplishing something positive are likely to perceive the change to be positive.

Avoidance needs underlie many of the beliefs associated with negative perceptions. Also important are power motives that are seen as undermined by impending changes. People who perceive that they will be net losers because of the change are likely to have negative perceptions of the change.

It is critical to note that most people will not express their perceptions, positive or negative, in terms of the needs and beliefs in Figure 7.3. Instead, they will argue about the necessity of changing, the possibility of changing, and the desirability of changing. Usually, it is difficult to convince people to change their perceptions. A more productive strategy is to work on the level of needs and beliefs.

Evaluate Hypotheses

I have no experimental data to support the hypotheses reflected in Figure 7.3. Instead, I can only provide anecdotal evidence. Besides my own company, which has undergone the types of change discussed in this section, I have helped numerous companies, both large and small, to consider the possibility of changing from defense to nondefense markets. In the process of running seminars and workshops related to this topic, I have repeatedly encountered the types of perceptions indicated in Figure 7.3.

After talking with many hundreds of people, in several tens of enterprises, about their positive and negative perceptions, the needs and beliefs in Figure 7.3 became apparent. The NBP Model offered a way to integrate these observations. Consequently, there is much informal evidence to support the hypothesized needs and beliefs. While this evidence lacks rigor, it is sufficient to serve as a basis for proceeding.

Modify Alternatives

In the context of this case study, modifying alternatives means changing the way in which the shift from defense to nondefense markets is represented. The necessity of change needs to be communicated and explained. The enterprise's inability to maintain the past should be emphasized. It is of particular importance that the benefits of the new future be extolled. In this way, the necessity of changing can become an opportunity for an even better future.

Painting this picture may be straightforward, especially for those who are leading the change and intimate with its underlying rationale. However, it may take a long time to educate everyone else. People need to be reassured that they have important roles to fill in the new enterprise. They need to be confident that they can fill these roles and, to the extent necessary, that training will be provided. Especially difficult may be the slow process of helping them to appreciate and value these new roles.

As noted in Chapter 3, acceptance can often be enhanced if continuity with the past is emphasized. By clarifying and explaining strong links to past competencies and technologies, people may come to see that the new future is an outgrowth of the past rather than an arbitrary exit off the old highway. To the extent that people can be convinced that growth and change are necessary, they are likely to accept natural evolution. They may come to see that their own growth and maturing are part of the same process.

The "vision" of the future should be clearly portrayed. As discussed earlier in this chapter, this vision should include both "outward" and "inward" views. I have found that many people are primarily concerned with the nature of their work, the working conditions, and the people with whom they work. A vision in these terms may be much more compelling than an abstract statement concerning quality and customer satisfaction. While such statements are important, they are not sufficiently compelling to allay concerns and transform negative perceptions to positive.

Modify Situations

For the longer term, change can be slowly facilitated by changing the stories of the enterprise. The change of markets can become part of the "epic" of the enterprise's growth and maturation. The ability to change and take advantage of new opportunities can be portrayed as one of the enterprise's greatest strengths.

Stories of constant change may be disconcerting for many people. Thus, it may be necessary to show some stability in the depiction of constant change. This is where the enterprise's values can be shown to provide the foundation for change. These values can reflect the ways in which people are treated, both as customers and employees. The role of the enterprise in the community and society is also part of the enterprise's values.

It also can be very important that the community and society show that they value the enterprise and its willingness to change and adapt to new markets. Public policies can be created to encourage, or at least not discourage, transitions from defense to nondefense markets. This might include, for instance, tax credits for training.

Plan Life Cycle

As noted in the discussion earlier in this chapter, the life cycle of enterprise changes are best considered as an evolving series of changes. Changes will proceed more smoothly if they are planned. However, it is impossible to anticipate all of the additional needs to change that will emerge. Hence, plans should not be portrayed as promises. Instead, plans should be viewed as, hopefully, well-articulated intentions to create particular changes and, in the process, identify the next changes necessary in the evolutionary path. The overall plan should be to create an enterprise that is well skilled in the process of change.

Summary

This case study has focused on the problem of changing from defense to nondefense markets. The difficulties of such changes were discussed. An important element of these difficulties is the reluctance of the organization to change. The NBP Model, Template, and Analysis Methodology were used to determine the underlying causes of this reluctance and identify ways to deal with these causes.

IMPLICATIONS OF ANALYSES

This chapter has considered two applications of the NBP Analysis Methodology to problems of organizational change. It is useful to consider the results of these analyses in terms of the constructs introduced in Chapter 2, namely, expectations, attributions, and mental models. In this way, the generality of these results is more apparent.

Expectations play a central role in both compelling and resisting change. Expectations compel change as people in all countries compete for scarce resources. They expect that their efforts will lead to improved standards of living and, therefore, are willing to work long hours for low wages, for example, in order to improve the lives of their families.

For those who are better off, expectations are that the "good life" will continue. The availability of high-paying jobs is taken as a given. It is assumed that a college education will guarantee a good job. "Delayering," "downsizing," and, more recently, "rightsizing" are phenomena that conflict with these expectations.

Thus, the expectation is that change will not be necessary. But recent evidence has become compelling. How do people react? My experiences are that people initially deny the necessity of change. They attribute the evident problems to outside sources and, in the process, often make a "fundamental attribution error," as discussed in Chapter 2.

To the extent that the necessity of change is recognized, a common strategy is to seek the potential benefits of change with the smallest actual change possible. People try to "muddle through." If they recognize that fundamental and substantial change may be necessary, they frequently conclude that their enterprise is not capable of such change.

If change is, nonetheless, attempted, those who perceive the plans negatively often try to undermine plans by asserting that the benefits sought will not be achieved. They attribute ulterior motives to the enterprise, as well as to various ntities outside the enterprise. Below the surface, however, strong fs are often influencing these perceptions, frequently without t.

nge is first pursued via single-loop learning. Attempts are made processes and fix problems within processes. In some cases, it is realized that mental models may be wrong. This realization may provide a starting point for double-loop learning.

While myths and models were once interpreted as facts, other possibilities may now be explored. The enterprise's espoused theories of actions may be compared to theories-in-use. Alternative mental models of the marketplace and the enterprise itself may be entertained.

New visions have to be created and maintained. These visions may include both somewhat abstract statements concerning the enterprise's relationships with its stakeholders, and more concrete pictures of the likely day-to-day life of the enterprise after changes have been successfully implemented.

The process of creating and maintaining these types of visions will, albeit often very slowly, lead to new stories of the enterprise. Successful stories of change involve changing the stories of success. In other words, changes will be successful to the extent that the mental models of the stakeholders of the enterprise change in a manner that reflects the enterprise's new ways to succeed.

The process of changing expectations, attributions, and mental models can be extremely time consuming. During this process, it is impossible for top management to overcommunicate. Top management has to lead in understanding the forces for change, avenues for changes, and strategies for change. The insights gained from an NBP analysis—the relationships identified among needs, beliefs, and perceptions—provide the catalysts for change.

SUMMARY

This chapter and the previous chapter (Chapters 6 and 7) have focused on understanding the marketplace and enabling the enterprise. The primary concern has been with relationships between a single organizational entity and its external stakeholders (Chapter 6), as well as its internal stakeholders (Chapter 7). The NBP Model, Template, and Analysis Methodology were used to organize our knowledge and identify avenues of change that are supportive of needs and beliefs.

In the next two chapters, we consider archetypical innovation problems that involve multiple organizational entities. Needs, beliefs, and perceptions associated with disputes and conflicts are discussed in the context of several case studies. These discussions are, admittedly, somewhat more speculative than earlier discussions. Nevertheless, they provide a stern and interesting test of the model, template, and methodology.

REFERENCES

Argyris, C. (1982). *Reasoning, learning, and action: Individual and organizational.* San Francisco: Jossey-Bass.

Argyris, C., and Schon, D. A. (1978). *Organizational learning: A theory of action perspective.* Reading, MA: Addison-Wesley.

Blanchard, R. E., and Blackwood, W. O. (1990). Change management process. In H. R. Booher (Ed.), *MANPRINT: An approach to systems integration* (Chapter 3). New York: Van Nostrand Reinhold.

Booher, H. R. (Ed) (1990). *MANPRINT: An approach to systems integration.* New York: Van Nostrand Reinhold.

Brown, J. S. (1991). Research that reinvents the corporation. *Harvard Business Review,* January–February, 102–111.

Brown, J. S., and Duguid, P. (1991). Organizational learning and communities-of-practice: Toward a unified view of working, learning, and innovation. *Organization Science, 2,* 40–57.

Conlon, E. J. (1983). Managing organizational change. In T. Connolly (Ed.), *Scientists, engineers, and organizations* (pp. 363–378). Monterey, CA: Wadsworth.

Institute of Electrical and Electronics Engineers (1992). Special issue on defense conversion. *IEEE Spectrum, 29,* (12).

Kanter, R. M. (1989). *When giants learn to dance.* New York: Simon and Schuster.

Lloyd, T. (1990). *The 'nice' company: Why 'nice' companies make more profits.* London: Bloomsbury.

McClelland, D. C. (1987). *Human motivation.* Cambridge, UK: Cambridge University Press.

Mokyr, J. (1990). *The lever of riches: Technological creativity and economic progress.* Oxford, UK: Oxford University Press.

Rouse, W. B. (1991). *Design for success: A human-centered approach to designing successful products and systems.* New York: Wiley.

Rouse, W. B. (1992). *Strategies for innovation: Creating successful products, systems and organizations.* New York: Wiley.

Rouse, W. B., and Boff, K. F. (Eds.) (1987). *System design: Behavioral perspectives on designers, tools, and organizations.* New York: North-Holland.

Senge, P. M. (1990). *The fifth discipline: The art and practice of the learning organization.* New York: Doubleday/Currency.

Thurow, L. (1992). *Head to head: The coming economic battle among Japan, Europe, and America.* New York: Morrow.

Wilkins, A. L. (1989). *Developing a corporate character: How to successfully change an organization without destroying it.* San Francisco: Jossey-Bass.

Chapter **8**

Settling Sociotechnical Disputes

Sociotechnical disputes usually involve multiple entities that disagree about the current state of affairs and/or planned course of events involving society and technology. The parties in these disputes may disagree about facts—for example, whether a waste product is toxic—and about priorities such as preserving jobs vs. focusing on environmental protection. They may also disagree about past facts and priorities, such as what was originally decided and whether the issues governing those decisions were appropriate.

Settling such disputes involves deciding who gets which resources, positions, and consequences. This presents a type of problem that was not central to earlier discussions. While Chapters 6–7 focused on what can usually be cast as win–win situations, this chapter and the next deal with what often become win–lose situations. One or more parties get the resources and positions, and other parties get the consequences. Not surprisingly, all parties in the dispute focus on becoming the "winners," rather than the "losers."

An important premise of this chapter, and the next, is that below the surface it might be possible to find win–win situations in terms of needs and beliefs. To a great extent, our primary concern is with what Lawrence Susskind and Jeffrey Cruikshank (1987) term "negotiated approaches to consensus building." This approach, they argue, enables transforming win–lose situations into win–win, or what they call "all-gain" situations.

Susskind and Cruikshank note that "most participants in public disputes are so accustomed to thinking in win–lose terms that they cannot imagine an approach that seeks to ensure mutual gain for all contending parties." Consequently, all parties focus on winning and deadlocks frequently result. While it is not easy to transform win–lose situations into win–win situations, such transformations are often necessary if expensive, time-consuming, and frustrating litigation is to be avoided.

A primary means to achieving this goal involves "efficient packaging of items that disputants value differently." To accomplish this, the scope of the negotiation often must be expanded. By linking outcomes of several issues of concern to the parties involved, it may be possible to transform win–lose situations into what Susskind and Cruikshank term "integrative bargaining opportunities." The key element of making integrative bargaining work is the availability of items that disputing parties value differently. If packaged appropriately, all parties can "win."

Howard Raiffa (1982) notes that "the potential of finding joint win-win situations depends on the exploitation of differences between beliefs, between probabilistic projections, between tradeoffs, between discount rates, and between risk preferences." This chapter employs the Needs–Beliefs–Perceptions (NBP) Model, Template, and Analysis Methodology to focus on these types of differences. The resulting deeper understanding of conflicts provides an important basis for the types of transformation necessary to creative resolution of sociotechnical disputes.

Before delving into approaches to negotiation, it is useful to consider the different levels at which disputes can be addressed. At one level—the surface—our concern is with conflicting perceptions of viability, acceptability, and validity. Various parties may disagree, for instance, about the severity of a toxic waste problem and the priorities for resolving this problem. If we stay at this level, we may become stuck.

However, below the surface, we can address this dispute more creatively. *Why* do the different parties have conflicting perceptions? A simple answer is that some parties have incorrect facts. If this is the sole cause of the differing perceptions, the dispute is probably readily resolvable. However, this is seldom the total cause of dispute.

Another type of problem hinges on a phenomenon discussed in Chapters 3–4 such as, interpreting myths and models as facts. It is quite common for people from different disciplines to implicitly assume their models to be "true." An example of this was illustrated in Chapter 4 in the context of disputes about nuclear power. Consequently, those who do not share the same myths or models are perceived to be "wrong." However, by understanding all of the belief systems involved, as well as the underlying needs, it may be possible to transform "right vs. wrong" into a more creative exploration of perceptions, values, and concerns as affected by differing needs and beliefs. This chapter pursues such an exploration.

As a final issue before pursuing discussions of types of negotiations, it is useful to note that many negotiations are not explicit. On the surface, there are arguments, debates, and disagreements. The climate may become heated. Very strong statements may be made, a few of which may later be regretted. However, if the implicit negotiation can be made explicit, it may be possible to move beyond fiery rhetoric and face the real issues. Thus, creative solutions may be more likely when disputes are recognized instead of submerged. Of course, the willingness to proceed in this direction may depend on being confident of making progress. This chapter provides a means for having this confidence.

APPROACHES TO NEGOTIATION

Not all negotiations are the same. Several characteristics distinguish among types of negotiation. Figure 8.1—adapted from the work of Howard Raiffa (1982)—summarizes many of the attributes of negotiations that differentiate types of negotiation situation.

A particularly key attribute concerns whether there are two or more parties. As noted in Chapter 4, with more than two parties things get much more complicated and fundamental limits can be encountered (Arrow, 1963). An additional concern is whether the parties are monolithic, or involve factions within parties. In the latter case, there can be negotiations within negotiations.

Another concern involves one-time decisions vs. repetitive decisions. It is only reasonable to play for "winning on the average" when the game is played many times. For situations involving one-time resolution of an issue, the decision becomes much more crucial.

- Are there more than two parties?
- Are the parties monolithic?
- Is the game repetitive?
- Are there linkage effects?
- Is there more than one issue?
- Is an agreement required?
- Is ratification required?
- Are threats possible?
- Are there time constraints or time-related costs?
- Are the contracts binding?
- Are the negotiations private or public?
- What are the group norms?
- Is third-party intervention possible?

FIGURE 8.1. Attributes of conflict situations (adapted from Raiffa, 1982).

Negotiations can involve more than one issue. In fact, as noted earlier, it may be easier if multiple issues are involved, especially if the different parties value the various issues in different ways. This provides opportunities for integrative bargaining.

Another set of concerns involves the need for formal agreements and ratification of agreements. These needs make negotiations much more complicated. The important issue, of course, is recognizing those situations where informal agreements, that is agreements in principle, are only binding when they become formal agreements, that is agreements in practice. In some cases, agreements may not be binding and renegotiation may be a continual necessity.

Raiffa (1982) also notes that negotiations may be affected by the possibility of threats, as well as time constraints and time-related costs. Such considerations may create pressure for settlements. At the very least, these issues may cause the nature of acceptable agreements to evolve as costs and other concerns accrue.

Public negotiations tend to be quite different than private negotiations. Positions taken and argued in public are subject to much greater scrutiny, for a wider range of reasons, than those made in private. As a result, the public agenda tends to be much broader than a private agenda focused on settling a particular dispute.

Another issue is group norms. This issue concerns whether disclosures and statements can be assumed to be true and reflect the actual priorities of the parties putting them forth. If, on the other hand, they only represent tactical moves in the negotiation game, the information in these statements is used somewhat differently. For example, if one cannot count on the truth of supposed facts provided by disputing parties, one must carefully consider the intent of each statement.

The final entry in Raiffa's list of questions concerns the possibility of third-party intervention. Many disputes are such that progress is difficult or unlikely if the disputing parties attempt to settle among themselves. Assisted negotiation is discussed later in this chapter.

It is useful to conclude this introduction to negotiation by considering the possibility of using traditional majority-rule democracy as a means for settling disputes. Susskind and Cruikshank (1987) discuss the shortcomings of this approach. They note the well-known tendencies toward a "tyranny of the majority." Further, political commitments are often short term, due to the need to get re-elected. However, the possible consequences underlying sociotechnical disputes are often quite long term, substantially beyond the planning horizon of most politicians.

The voting process also has its problems. Voting is usually not timely, typically every two to four years. Consequently, issues often cannot be resolved when needs and opportunities arise, Further, the choice offered is usually quite simple—"yes" or "no." Moreover, voting by representatives is often subject to influence by lobbyists.

The usual voting processes may be much too simple for the problems at hand.

The complexities of environmental protection, economic development, and trade imbalances, for instance, are not well addressed by simple yes/no voting. With its typical emphasis on winner-take-all solutions, the usual process does not lend itself to exploring creative possibilities for win–win resolutions.

Nature of Negotiation

Susskind and Cruikshank (1987) discuss what they feel are the characteristics of good negotiated settlements. First of all, the settlement should be viewed as fair, which usually means that the process whereby the settlement was reached is viewed to be fair. They note that "in a public dispute, a good process produces a good outcome; and a better process, a better outcome." Elisabeth Pate (1983) in her studies of public acceptance of risk reaches similar conclusions and says that "acceptable decision processes are necessary for acceptable risks."

Susskind and Cruikshank also suggest four tests of fairness. They first ask if all stakeholders were given a chance to be involved. Their second concern is whether there were opportunities for systematic review and improvement of evolving options. The third test is whether the negotiation process is perceived as legitimate after it has ended. Finally, one must ask if a good precedent is set by the results in the eyes of the community.

Good settlements should also be efficient in terms of time and cost. Such settlements also score high in wisdom in the sense that if both sides cooperate there will be minimal chances of being wrong. Good settlements are also stable in terms of their endurance. Put succinctly, good settlements do not dictate a particular agreement—they require a particular type of process.

Susskind and Cruikshank argue that consensual approaches to negotiation are more likely to be subsequently judged as good. However, they note that "consensual solutions are better—and will be accepted—only if all stakeholding parties are confident they will get more from a negotiated agreement than they would from a unilateral action, or from conventional means for resolving distributional disputes." Therefore, in making consensual agreements the preferred approach requires that the best alternative to a negotiated agreement not be likely to yield better results.

Creating such perceptions can be difficult, especially when strategic misrepresentation is common. As Howard Raiffa (1982) notes, "The art of compromise centers on the willingness to give up something in order to get something else in return. Successful artists get more than they give up. A common ploy is to exaggerate the importance of what one is giving up and to minimize the importance of what one gets in return. Such posturing is part of the game."

An important element of avoiding these aspects of "the game" involves getting people to communicate in a collaborative manner. Raiffa offers several tactics that can help. One tactic focuses on getting people involved in developing a model of the phenomena of concern. Another tactic employs joint, cross-organizational problem-solving teams. Raiffa suggests that negotiating parties adopt "codes of conduct" that require communication. Based on his extensive experience with cross-country cooperation, he further suggests that ongoing institutes that foster long-term cooperation can provide a means for continual communication, rather than being limited to communication in dispute situations.

Raiffa discusses communication difficulties associated with many-party negotiations. In such situations, he argues for a "single negotiating text" whereby a neutral party prepares drafts that the negotiating parties critique. He also considers difficulties with bilateral negotiations which, he notes, usually require three agreements—one across the table and one on each side of the table.

There are a variety of ways in which negotiations can get off track. Susskind and Cruikshank (1987) discuss the various ways that "psychological wars" can emerge, leading to escalation and parties being trapped in their positions. The phenomena discussed in Chapter 2—expectations, attributions, and mental models—can lead to self-fulfilling prophecies, selective perceptions, and attributional distortions. Raiffa (1982) discusses the impact of impatience in negotiations, which usually leads to less satisfactory results for the most impatient.

Clearly, negotiations can be very complicated. Many psychological phenomena are often involved and it is easy to become stalemated. By recognizing this possibility and adopting an explicit negotiation process, the impact of these types of problem can be controlled.

Negotiation Processes

Susskind and Cruikshank (1987) contrast two basic approaches to negotiation: conventional and consensual. Figure 8.2 compares these two techniques. These authors argue, quite convincingly, that the consensual approach should be preferred, as it provides a means for breaking the typical impasses that arise during disputes.

Susskind and Cruikshank suggest several prerequisites for negotiating using this approach. First, the key players must be identified and convinced that it is in their interest to negotiate in this manner. To obtain their agreement it is often necessary that power relationships be balanced, with no party being able to impose its will on the other parties. For each participating party, a legitimate spokesperson has to be identified and recruited. It is also important that any deadlines be realistic. Finally, it is critical that disputes be framed such that agreements do not require compromising any sacrosanct values.

Additional issues for citizen groups concern the availability of resources to participate and the degree of consensus within the group. For business groups, issues also include the extent to which a mandate exits to proceed, whether someone with relevant negotiating experience is available, and the potential for

ATTRIBUTES	CONVENTIONAL APPROACHES	CONSENSUAL APPROACHES
Outcomes	Win–lose; impaired relationships	Win–win; improved relationships
Participation	Mandatory	Voluntary
Style of interaction	Indirect (through lawyers or hired advocates)	Direct (parties deal face-to-face)
Procedures	Same ground rules and procedures apply in all cases	New ground rules and procedures designed for each case
Methods of reaching closure	Imposition of a final determination by a judge or official	Voluntary acceptance of final decision by the parties
Role of intermediaries	Unassisted; no role for intermediaries	Assisted or unassisted; various roles for intermediaries
Cost	Low to moderate in the short term; potentially very high in the long term	Moderate to high in the short term; low in the long term if successful
Representation	General-purpose elected or appointed officials	Ad hoc; specially selected for each negotiation

FIGURE 8.2. Comparison of approaches to resolving disputes (Susskind and Cruikshank, 1987).

patience with deadlines. A primary business concern is also the intention, or lack of intention, to continue doing business in the same community.

Raiffa (1982) also suggests a set of issues that negotiators should address. To prepare for negotiations, you should know where you stand and your aspiration levels. You should also gain as much knowledge as possible about the other parties. Negotiating conventions and the logistics of the negotiations should be settled. Role playing can be used to prepare for negotiations.

During negotiations, he suggests that it is essential to control your reactions to the first offer, particularly if it is extreme. Throughout the pattern of concessions, care should be taken to preserve the integrity of your party's position. At the same time, you should be open to reassessing perceptions of what is achievable and what is reasonable.

As negotiations move to completion, you should be aware of what commitments are being made. Commitments that must be broken should be done so gracefully. Similarly, you should help your adversaries to break their commitments gracefully. In this process, it may necessary to introduce a neutral third party. It also may be necessary to broaden the domain of the negotiation in order to achieve agreement. This is especially true for consensual negotiations.

Susskind and Cruikshank (1987) outline a process for consensus building. Prior to negotiations, the focus should be on identifying and selecting representatives of all stakeholder groups, drafting negotiation protocols, and setting the agenda. They suggest joint fact-finding, in order to establish communication and trust among the disputing parties.

Negotiations involve inventing and packaging options that will be perceived as win–win. In this invention and packaging process, Susskind and Cruikshank argue that parties must focus on their concrete interests, rather than their abstract positions. Numerous iterations may be required to arrive at acceptable written agreements. Representatives of the negotiating parties then return to their parties to seek ratification of the agreement. If agreements are reached, parties are asked to bind themselves to these commitments.

In some cases, the resulting agreements may be informal and, subsequently, have to be linked to formal decision-making processes such as, for example, voting. Beyond formalizing agreements, means should also be created for monitoring the terms of agreements. This should also include consideration of how changing contexts will enable renegotiation.

The processes outlined thus far can proceed in either an unassisted or assisted manner—in other words, without or with neutral third-party assistance. Susskind and Cruikshank suggest three preconditions for unassisted negotiation. First, the issues in dispute, as well as the array of stakeholding parties, should be relatively few in number and readily identifiable. Second, the stakeholders must be able to establish sufficient channels of communication to permit joint problem solving. Finally, the uncertainty surrounding the outcome of unilateral action must be moderately high for all stakeholders.

If one or more of these preconditions are not satisfied, assisted negotiation is likely to be of value. For consensual negotiations, three forms of assisted negotiation include facilitation, mediation, and nonbinding arbitration. Facilitators merely orchestrate the discussion process. Mediators are more active, meeting separately with each of the disputing parties. Arbitrators also create packaged agreements themselves which are submitted to the parties for review. If the arbitration is nonbinding, the parties are not required to accept this agreement. However, any new or revised agreements are created by the arbitrator.

Raiffa (1982) suggests several ways in which third-party intervention can help. Third parties can bring the disputing parties together, establish a constructive atmosphere for negotiation, and collect and judiciously communicate selected confidential information. Third parties can also help the parties clarify their interests, revise unreasonable claims, and loosen commitments that are inhibiting progress. They also can seek and create win–win packages, keep negotiations going, and articulate the rationale for agreements.

Susskind and Cruikshank (1987), as well as Raiffa, outline the tasks of mediators. In both of their compilations, mediators are depicted as convening the negotiations, keeping the parties on track within the process chosen, recording minutes and agreements, packaging options, and providing a degree of stability and control that is difficult to achieve without third-party assistance. All of these authors agree that third-party assistance can be the key to avoiding stalemate.

Summary

This section has summarized the state of knowledge on approaches to negotiation. The nature of negotiation was elaborated and negotiation processes were discussed. Clearly, there are successful approaches for settling sociotechnical disputes, as the many examples in Raiffa (1982) and Susskind and Cruikshank (1987) illustrate. In the remainder of this chapter, we examine the potential roles of the NBP Model, Template, and Analysis Methodology within negotiations processes.

ENVIRONMENTAL PROTECTION vs. ECONOMIC DEVELOPMENT

One type of sociotechnical dispute that is likely to be increasingly prevalent concerns *apparent* trade-offs between environmental protection and economic development. The word "apparent" is emphasized because it is not at all clear, as later discussion illustrates, that one of these desires has to be traded for the other. Nevertheless, the public debate is usually cast in these terms.

Stakeholders in these disputes are numerous. Communities, industry, and government are central. Other groups such as Greenpeace also play significant roles. All of these stakeholders want the environment protected. At the same time, there is a tendency to want others to pay for it.

An experience that I had a few years ago provides a good illustration of the nature of the conflicts that emerge. I was a member of a committee established to oversee the destruction of the United States stockpile of unitary chemical weapons. In contrast to binary weapons, where lethal mixtures are not created until they are used, unitary weapons are stored in a form that provides the potential for deadly and disastrous releases into local populations in the event of accidents.

The effort being overseen by this committee involved elimination of these weapons at several sites in the continental United States and one offshore. No one disagreed about wanting these weapons destroyed. However, there was much discord about the best means to accomplish this goal.

The communities in which these sites were located wanted the weapons removed, perhaps to a remote site for destruction. However, the residents of surrounding communities did not want the weapons moved through their communities. Lawsuits were filed to force both possibilities, that is, move them and not move them. Greenpeace also became involved; it was especially concerned with the "remote" sites where the weapons would be moved and destroyed.

It was finally decided that it would be safer to destroy the weapons at their storage sites. This led to fears that the incinerators that would be built would subsequently be used for other toxic wastes. Communities became concerned that availability of these incinerators would be used to justify designating their communities as permanent toxic waste facilities (Smothers, 1992).

Further complicating this dispute was the lucrative construction contracts for building these incineration facilities. Many jobs would be provided. As a result, communities both wanted and did not want these facilities. Moreover, a variety of contractors vied for these contracts, occasionally with a little help from influential politicians.

This example typifies many aspects of sociotechnical disputes involving environmental protection vs. economic development. On the one hand, there are disputes about facts—for example, the real risks of wastes and chemical weapons. There are also disputes about priorities—for instance, a large number of construction jobs now vs. ambiguous consequences later. In addition, benefit-cost tradeoffs are not common across all stakeholders. Finally, all concerns cannot be fully satisfied. You cannot both move and not move the risky materials.

James Flynn, Roger Kasperson, Howard Kunreuther, and Paul Slovic (1992) consider citing of nuclear waste facilities. They note that social acceptability has always taken a backseat to technical concerns. However, the public dreads all things nuclear as well as the stigma of being the "nuclear dump state." Nonetheless, Congress and industry have been unwilling to put up with the drawn-out process of involving the public. The result has been public opposition and distrust.

Since safety cannot be 100 percent guaranteed, Flynn and his colleagues argue that the public should decide what is acceptable. They note that Europeans have progressed more smoothly because they have not tried to solve this problem too quickly and "have placed considerations of equity, fairness, and social acceptability on an equal footing with technical goals." To have any prospect of success, these authors assert that the United States also must develop approaches that are "socially acceptable as well as technically sound, collaborative rather than preemptive, and predicated on persuasion and negotiation rather than coercion."

Flynn and co-workers (1992) argue for a voluntary siting process, with appropriate compensation to the area chosen. Raiffa (1982) discusses the Massachusetts law regarding siting of hazardous waste facilities that requires community participation in decisions and compensation for local damages. Elisabeth Pate's (1983) perspective is similar in her assertion that an acceptable decision process will reduce the time of conflict, decision making, and implementation; increase economic efficiency; and lead to wider acceptance of results based on participation in the decision by those who are truly affected. She argues that conflict resolution and compensation procedures should be central.

Considering the perceptions of the various parties in these types of dispute, Richard Barke and Hank Jenkins-Smith (1992) compare perceptions of over 1,000 physical and life scientists, and those of academia and government/industry, with respect to environmental risks, particularly regarding nuclear wastes. Life scientists tend to perceive the greatest risks from nuclear energy and nuclear waste management, are most skeptical about the effects of technology on society, perceive higher levels of overall environmental risk, strongly oppose imposing risks on unconsenting persons, and prefer stronger requirements for environmental management. Independent of field of research, those in academia and state and local government tend to perceive risks to be higher than those who work in firms of business consultants, federal government, and private research laboratories. Clearly, all scientists do not agree—it depends on both discipline and organizational culture.

Lester Thurow (1992) contrasts economists and environmentalists: "Economists focus on goods and services as the primary focus of attention, with environmentalism as a secondary issue. Environmentalists see the world exactly in reverse. A clean environment is primary; more goods and services are secondary." To an economist, conservation is impossible "unless people today are being inefficient and irrational in their behavior—a possibility that environmentalists believe and economists reject." The conclusions of Barke and Jenkins-Smith (1992), as well as those of Thurow, serve to illustrate how beliefs may influence perceptions.

One of the strongest beliefs affecting these types of debate concerns the aforementioned *apparent* trade-off between the environment and the economy. Curtis Moore (1992) discusses relationships among jobs and the environment. He argues that environmental problems are also economic opportunities, as Japan and Germany have shown for pollution-cutting technologies, including reduction of pollutants and recycling. Moore notes that Japan and Germany now lead the United States in many pollution-control technologies that were invented in the United States but not commercialized due to lack of incentives. It is quite likely that our mental model of the relationships among environmental and economic issues will hinder us from seeing an opportunity rather than a crisis.

The remainder of this section is devoted to applying the NBP Analysis Methodology in order to probe below the surface of disputes involving the environment and economy. Rather than concentrate on one particular dispute, this example focuses on an archetypical dispute involving communities and environmentalists on one side, and industry and employees on the other. The goal of this analysis is to illustrate the roles that needs and beliefs can play in conflicting perceptions.

Identify Measures

Measures of viability reflect benefits and costs. From the point of view of communities and environmentalists, benefits of environmental protection include clean air and water which imply reduced risks of diseases, as well as avoidance of large cleanup costs in the future as problems worsen. The costs are potential loss of jobs and perhaps whole companies and, consequently, the decline of local economies and loss of tax revenues.

From the perspective of industry and employees, benefits are similar in nature but tend to be seen as smaller in magnitude. Costs, on the other hand, are viewed to be enormous. These costs do not represent job loss in general to employees—they portend loss of particular jobs, namely, theirs. For industry, the money comes out of its pockets, not the generic "economy."

Acceptability hinges on how much change is necessary to meet standards vs. how much change is possible within company resources. Communities and environmentalists are likely to argue for no compromises on absolute levels of pollutants or wastes. Companies and employees tend to counter with more modest and incremental improvements that can be accomplished without overly straining a company's infrastructure.

Validity for communities and environmentalists relates to the extent to which companies' plans will meet standards in the required time frame. From the perspective of companies and employees, validity is seen much more broadly in terms of the effects on the overall production process, whose products include much more than pollutants and waste. The business has to remain profitable, via its products, while also dealing with the environmental problems at hand.

Determine Alternatives

For the sake of illustration, the alternative of interest in this case study is the requirement that a group of companies develops and implements plans to meet environmental standards regarding pollutants and wastes. These pollutants could be the products of combustion or chemical processing of materials. The wastes might be nuclear or chemical wastes. The companies might be located in any place where such problems could emerge. The analysis that follows, therefore, is quite general and serves to illustrate the possibilities provided by an NBP analysis.

Assess Perceptions

From the point of view of communities and environmentalists, validity will be perceived as high if companies' plans outline an auditable set of steps that will meet all standards as quickly as is technically possible. Anything less may be suspect.

However, from the perspective of companies and employees, such a plan may be seen as lacking in validity because it does not take realistic account of companies' resources and the need to remain productive in the process of meeting standards. The companies and employees may also question the standards and the projected environmental consequences of not meeting standards.

To a great extent, disputes about standards and consequences may seem to be the most straightforward. Studies can be performed to discern disputed facts, and mutually agreed-upon models can be used to perform projections. This requires, of course, that all parties have the same beliefs about the abilities of science and technology to uncover facts and formulate appropriate models. It further requires that all parties trust each other regarding information provided concerning, for example, companies' production processes.

Acceptability is inherently more complicated. Perceptions are strongly related to perceived risks and whether these risks can be tolerated. Communities and environmentalists may perceive long-term risks as uncertain and, hence, large and intolerable. Companies and employees are likely to see short-term risks as very large, as well as much more concrete than long-term risks. Other issues that can affect perceptions of acceptability include concerns about fairness, saving face, preserving positions, and hidden agendas such as, for instance, upcoming elections.

Perceptions of viability are likely to be influenced by expectations. For example, if those on one side of the issue expect a complete loss, any gain is likely to be viewed as viable. In contrast, those who expect complete victory are likely to find anything else lacking in viability. As noted in earlier discussions, negotiations are usually most fruitful when the disputing parties feel that they will gain by negotiating a mutually agreeable compromise. From this perspective, both sides' perceptions of viability are likely to be influenced by what short-term and long-term losses and gains are packaged in the proposed agreement.

Define NBP Templates

For this analysis, we will employ one template for communities and environmentalists (Fig. 8.3) and one for industry and employees (Fig. 8.4). It must be noted that a realistic analysis of a particular dispute would likely require multiple templates, including at least one for government agencies. However, this case study would be overly complex if such complications were considered.

Hypothesize Beliefs

The beliefs columns of Figures 8.3 and 8.4 were completed by first reviewing the earlier material in this chapter and then reviewing Figures 3.6, 3.11, 4.3, and 6.5. As usual, entries were reworded to fit the context of the case study.

It is useful to contrast the nature of the entries in these two figures. For the most part, communities and environmentalists' beliefs focus on ends, such as, meeting

ATTRIBUTE	PERCEPTIONS	BELIEFS	NEEDS
Viability	Positive	Long-term benefits substantial and short-term costs reasonable	Needs to predict, be in control, and achieve stability
	Negative	Insufficient benefits and/or too slow, costs too high	Needs to deal with unknowns, uncertainties, and risks
Acceptability	Positive	Necessary standards met quickly	Needs to predict, be in control, and achieve stability
	Negative	Standards not met and/or too slow	Needs to deal with unknowns, uncertainties, and risks
Validity	Positive	Plan will work and achieve environmental goals, organization is capable	Needs to predict, be in control, and achieve stability
	Negative	Plan will not work, facts and models are wrong, organization is not capable	Needs to deal with unknowns, uncertainties, and risks

FIGURE 8.3. NBP Template for communities and environmentalists.

standards and gaining long-term benefits. Industry and employees, on the other hand, are more concerned with the balance between short-term and long-term gains and losses, as well as the means to achieve these ends.

Hypothesize Needs

The needs columns of Figures 8.3 and 8.4 also were filled in by first reviewing the foregoing discussion of the problem and then reviewing the entries in Figures 3.5, 3.7, 3.10, 4.2, and 6.4. Wording was modified as appropriate.

ATTRIBUTE	PERCEPTIONS	BELIEFS	NEEDS
Viability	Positive	Short-term and long-term gains and losses balanced, projected bottom line acceptable	Needs to recoup investments and make profits
	Negative	Short-term losses excessive and long-term gains unknown	Needs to deal with risks and hedge against uncertainties
Acceptability	Positive	Profile of changes agreeable in magnitude and timing	Needs of continuity, stability, prediction, and control
	Negative	Profile of changes too great and/or too quick	Needs to deal with risks and instability and hedge against uncertainties
Validity	Positive	Plan will work due to organization's capabilities and achieve both environmental and economic goals. Models and technology are sound	Needs to comply, defend choices, predict, control, and exercise skills
	Negative	Plan will not work economically, standards and projections wrong, models not sound, solutions NIH and NSH ("not invented here" and "not supported here," respectively)	Needs to deal with unknown, uncertainties, and risks

FIGURE 8.4. NBP Template for industry and employees.

Many of the needs are shared by both templates. Entries dealing with unknowns, uncertainties, and risks appear several times. Needs to predict, be in control, and assure stability are also common to both templates. Primary differences are the needs of industry and employees to recoup investments and make profits while also complying with social dictates and defending choices. Industry and employees also need to exercise their own skills to achieve the desired ends, while communities and environmentalists must depend on the capabilities of others.

Evaluate Hypotheses

The hypothesized beliefs and needs have not been evaluated in any formal manner. They primarily reflect a compilation of my personal experiences, leavened by the NBP Model, Template, and Analysis Methodology. Therefore, my assertions are perhaps best described as well-informed speculations.

For any real analysis of a specific environmental dispute, I would expect that evaluation of hypotheses concerning underlying beliefs and needs would be part of the negotiation process, perhaps part of the joint fact-finding advocated by Raiffa (1982) and Susskind and Cruikshank (1987). The methods and tools described in *Design for Success* (Rouse, 1991) might be employed in this evaluation.

Modify Alternatives

The problem outlined thus far seems to satisfy all of the requirements for pursuing negotiations in general, and consensual negotiations in particular. Both sides have much to lose by employing conventional win–lose approaches. Furthermore, the entries in their respective templates (Fig. 8.3 and 8.4) appear to indicate clearly that different aspects of a potential agreement are valued differentially.

Consequently, it appears that broadening the scope of the negotiations could help. Hence, beyond levels of pollutants and wastes, the negotiations should also include consideration of investments, profits, and jobs. Industry and employees are clearly concerned about their abilities to absorb short-term losses due to costs, disruption, and organizational changes that might be necessary.

The two sides might proceed by first agreeing on the negotiation process and the rules of the game which will be observed regardless of conflicting positions. This could then be followed by joint fact-finding of two forms. One effort would focus on the technical basis for the standards of concern, as well as projections of consequences if standards are not met. While this effort is unlikely to eliminate differences between parties in terms of what risks are acceptable, it should at least decrease differences about the nature of the risks.

A second fact-finding effort could emphasize gaining an understanding of the impact of compliance on companies' productivity and profitability. Projections of such metrics can probably be transformed into an assessment of the impact on jobs, accessibility of investment capital, and possible consequences for the companies' future well-being. Problems in these areas might be ameliorated by tax incentives, tax credits, wage concessions, and perhaps even bond issues.

Thus, the packaging of the eventual agreement could be quite broad, involving consideration of the companies' roles in the communities affected and the need to create bearable short-term costs in order to realize long-term gains. Differing beliefs and needs provide at least a partial basis for understanding the necessary components of the resulting package.

Modify Situations

To a great extent, the severity of such disputes is affected by how late problems are identified and what relationships exist prior to the dispute emerging. Closer ongoing cooperation among all of the parties involved can provide a strong basis for consensual negotiations when they become necessary. To this end, a broader sense of community is likely to be needed. As Scott Peck (1987) notes, this broader sense of community can help to defuse many types of "wars."

A fundamental problem in achieving such broadened cooperation may be the beliefs of the different parties about the respective roles of government and the marketplace. Cooperation implies, to some extent, a trading of complete freedom for mutual benefit. The various parties in a dispute may see this issue in very different ways. This potential problem could easily affect disputes involving the environment and the economy. However, consideration of this problem is delayed until later in this chapter when we discuss trade disputes.

Plan Life Cycle

As noted during the earlier discussions of approaches to negotiation, many disputes are not settled once and for all. Compliance may have to be monitored. And in consensual agreements in particular, there should be means to determine whether or not the evolving conditions of the agreement are still fair and reasonable.

From this perspective, the design of an agreement may be best viewed as an evolutionary design process. Measures and measurements of viability, acceptability, and validity should, therefore, be regularly reviewed, refined, and updated. The framework of *Design for Success* (Rouse, 1991) provides a means for planning the life cycle of such design processes.

Summary

This section has briefly described an analysis of disputes involving the environment and the economy. It was shown that use of the NBP Model, Template, and

Analysis Methodology can provide important insights that are useful within the broad negotiating frameworks discussed earlier in this chapter. Such insights can serve as catalysts for settling sociotechnical disputes.

TRADE DISPUTES

Trade disputes have long been common. As I was writing this chapter a good example emerged. The United States announced plans to impose 200 percent tariffs on wine imported from Europe to protest apparent protectionist policies of European countries—France especially—toward agricultural imports. The United States claimed that it wanted a "level playing field" for international trade.

This sounds straightforward. However, the underlying basis for such disputes easily becomes quite complicated, as this chapter illustrates. These complications are illustrated in the context of trade imbalances with Japan. In this discussion, the issues will be simplified substantially to include only two sets of stakeholders—United States industry and Japanese industry. While this distinction is no longer as crisp as it once was, it is sufficient for demonstrating how an NBP analysis can be applied to this type of dispute.

In contrast to the earlier case study in this chapter (i.e., environment vs. economy), trade disputes are less concerned about facts and current priorities. The nature and level of trade imbalances are usually quite clear. Further, it is often straightforward to relate such imbalances to past differences in priorities such as exemplified by differences between American and Japanese automobile companies. Accordingly, the state of imbalances and their causes may be obvious to all parties in the dispute.

The dispute, in essence, concerns what to do about the imbalances. Should the Japanese simply buy more American automobiles, even if they do not want them? Or do they want them, but Japanese policies make buying them prohibitive? Should, perhaps, the Japanese import huge amounts of rice from the United States to offset all of the cars that we purchase from them? However, the Japanese have consciously decided to keep their small farms in business, apparently because of the role of rice farming in their culture. Also, my guess is that automobile companies and workers would not particularly favor the trade of American rice for Japanese cars.

With many of the alternatives that are proposed, we quickly encounter cultural differences that contribute, at least in part, to the current situation. My experiences in Japan, and elsewhere in the East, have prompted this hypothesis. My impression is that people in these countries are extremely entrepreneurial and hardworking. They are family oriented, respectful of their elders, generous, and hospitable. Moreover, they are aesthetically oriented and, as indicated in Chapter 3, recognize and appreciate the subjective aspects of life in general and business in particular.

Also noted in Chapter 3 was Geert Hofstede's (1980) analysis of business cultures, one conclusion of which was that Japanese business culture is very masculine in nature.

While these aspects of Japanese culture may have some impact on their competitiveness relative to the United States, my perception is that the sources of their trade advantages are much more specific. Recall the discussion in Chapter 3 of Lester Thurow's (1992) distinction between individualistic capitalism in the United States and communitarian capitalism in Japan (see Fig. 3.12). Thurow argues that the overriding goal of American companies is to make money, while the primary goal of Japanese companies is to build a community. Typifying this difference is the emphasis in the United States on individual brilliance and the emphasis in Japan on skilled teams.

Evidence in support of this assertion is ample. For instance, Thurow notes that half of United States companies require profits within three years of investing, while only 1 in 10 Japanese companies has similar requirements. Hence, Japan has much more "patient capital." In addition, their interest rates—the cost of capital—are much lower. This issue is discussed in Chapter 12 of *Strategies for Innovation* (Rouse, 1992).

Citing statistics on R&D expenditures in 23 developed countries, Thurow says that the United States is 5th if total R&D is considered, 10th if military R&D is excluded, and 20th if all government-funded R&D is excluded. Clearly, much of United States expenditures in R&D is linked to defense missions. This is not the case in Japan and Europe, where economic concerns more heavily influence R&D investments.

Japan and European countries tend to have explicit industrial policies that clarify priorities and articulate agendas for technology investments. The United States, in contrast, has avoided establishing industrial policies, supposedly leaving priorities and choices to the free market. However, the United States has long had a de facto industrial policy that heavily emphasizes investments in defense R&D. Such investments may produce jobs, but they do not enhance competitiveness. Issues related to industrial/technology policies are also discussed in Chapter 12 of *Strategies for Innovation* (Rouse, 1992).

Trade imbalances are likely to be affected more positively by investments in industry and economic development than weapon systems. United States efforts in this direction are, for the most part, relatively recent. Philip Shapira, David Roessner, and Richard Barke (1992) discuss federal–state collaboration in industrial modernization. They discuss the 1988 Omnibus Trade and Competitiveness Act which mandated the National Institute of Standards and Technology (NIST) to take a greater role in diffusing new manufacturing technologies. The Act authorized NIST to establish regional Manufacturing Technology Centers (MTCs) to transfer manufacturing technology and techniques and disseminate scientific, engineering, technical, and management information to manufacturing companies.

Discussing the results of their evaluation, Shapira and his colleagues argue that the level of funding is much too low to be effective, and that this program should be expanded to an industrial extension service, similar to the agricultural extension service which was established in 1914. They note that the agricultural extension service has a budget of $1.3 billion, one-third of which comes from the federal government and two-thirds of which is contributed by state governments. In contrast, in the same period $79 million was spent on supporting manufacturing by all sources—state, federal, university, and industry.

In a separate study, Philip Shapira (1992) considers how Japan helps small manufacturers. One way is with 170 Kohsetsushi centers, described below, which serve 870,000 firms with fewer than 300 employees. (In the United States, there are 355,000 firms with fewer than 500 employees.) Comparing firms of this size in Japan and the United States indicates that Japanese firms are 50 percent more likely to employ numerically or computer-controlled machine tools and 400 percent as likely to use advanced machining centers and material handling robots. Work-force training is also more prevalent in Japan. Small firms there also have much closer and more positive relations with the big firms that they supply. Kohsetsushi centers play key roles in helping small companies adopt new technologies and train their employees. Many of these centers were founded in the 1920s and 1930s, modeled in part on the United States agricultural extension program. The total budget for the Kohsetsushi centers is $500 million per year.

In a recent visit to the United Kingdom, I talked with a variety of people from government and industry who are involved in supporting small- and mid-sized companies so they can become or remain competitive. My impression is that such companies are not yet models of efficiency and effectiveness. However, it is quite apparent what their intentions are. They are focused squarely on export markets and job creation.

The contrasts discussed thus far in this section imply that the United States and its competitors—Japan and Europe—have very different beliefs about the roles of government and the goals of business. In the rest of this section, the impacts of these differences and others are considered as possible explanations for the basis of our ongoing trade disputes.

Identify Measures

Measures of viability are concerned with benefits and costs. In the context of trade disputes, benefits include goods and services exported, and consequently jobs created. Costs are the resources associated with creating competitive goods and services, including costs of infrastructure, capital equipment, and training.

Acceptability relates to the barriers and avenues for change. Trade agreements that require cultural changes are likely to encounter acceptability problems. Similarly, agreements that cause valued domestic industries to disappear are likely to be perceived negatively in terms of acceptability. For example, agreements that would, in effect, eliminate the semiconductor industry in the United States would be viewed negatively regardless of increased exports in other areas that the United States would obtain in such hypothetical agreements.

Validity is concerned with the extent to which an agreement would solve the trade imbalance prompting the dispute. It is easy to imagine agreements lowering tariffs, for instance, but not necessarily increasing exports from the countries with negative balances because goods and services of these countries are not valued by other countries.

Determine Alternatives

Within the scope of this chapter, it is not possible to consider a particular trade agreement. However, in order to illustrate an NBP analysis of this archetypical innovation problem, a hypothetical trade agreement between the United States and Japan can be postulated. The analysis focuses on potential positive and negative perceptions of such an agreement, as well as the likely needs and beliefs underlying these perceptions.

Assess Perceptions

American industry is likely to perceive an agreement as valid if it "levels the playing field" by eliminating trade barriers in both directions. Of course, level playing fields do not necessarily even the score, especially if one team is much better than the other. Therefore, a high-wage, low-skill country will have great difficulty competing with a low-wage, high-skill country, regardless of the elimination of trade barriers. In theory, the wage differential will eventually decrease, but this will not solve the problems of a low-skill country.

Japanese industry appears to understand this principle, as does Germany, for example. In addition, time horizons in Japan tend to be longer. Consequently, perceptions of validity also focus on likely long-term trends rather than solely short-terms effects of agreements on imbalances. In fact, it can be argued quite reasonably that Japan's longer-term perspective is an important component in its very large trade surpluses.

As noted earlier, perceptions of acceptability are likely to be strongly affected by what is gained and what is given up. Gaining possibilities of exports and jobs in valued industries clearly enhances perceptions of acceptability. Having to trade positions in existing valued industries, or having to forego possible positions in industries that would be valued, undermines perceptions of acceptability.

Agreements whose success may depend on cultural changes could also affect perceptions of acceptability. The United States penchant for not supporting industry—with the exception of the defense industry—is deeply rooted in beliefs about the importance of free markets. Trade agreements whose validity hinges on government support for the affected industry are likely to encounter acceptability problems.

Cultural changes that might be accelerated by, as noted earlier, lowering barriers to importing rice into Japan clearly affect acceptability. Agreements that might threaten, even in a small way, the culture of individualism in the United States would face acceptability barriers. Agreements perceived to be unfair to one or more stakeholders, for any of a number of reasons, would likely lead to acceptability problems.

Perceptions of viability would reflect anticipated levels of exports and jobs resulting from an agreement. United States industry would tend to calculate benefits over a 3-5-year time horizon, while the calculation for Japanese industry would employ a horizon of 10 years or longer. Perceptions of benefits are likely to be strongly affected by the results expected if an agreement is not reached.

On the cost side of viability, United States industry would include perceptions of changes necessary to take advantage of export opportunities. As noted earlier, it is quite possible that the extent of these changes would be underestimated. For Japanese industry, costs would likely be tied to perceptions of what would be given up by reaching an agreement. These perceptions would be heavily affected by anticipated losses in the absence of an agreement.

Define NBP Templates

For the purposes of illustration, this analysis only considers two templates—one for the United States (Fig. 8.5) and one for Japan (Fig. 8.6). In an actual analysis involving a real dispute, there could be multiple templates, including one or more for each of the governments involved. Consequently, the analysis discussed in the remainder of this section is necessarily simplified.

Hypothesize Beliefs

The entries in the beliefs columns of Figures 8.5 and 8.6 were drawn from the discussion so far in this section, as well as from reviewing the entries in Figures 3.6, 3.11, 4.3, and 6.5. Wording was changed as necessary to fit the context.

There are several interesting contrasts between the two beliefs columns. The United States' beliefs focus on winning the competition via individual efforts involving homegrown solutions. Japan, on the other hand, focuses on delighting the marketplace through value and quality, adopting whatever solutions best accomplish these goals, and pursuing these goals as a long-term community-oriented effort.

There are clearly conflicting beliefs about the roles of individuals vs. the roles of organizations. In the United States, individuals who single-handedly best the competition are extolled. In Japan, teamwork involving multiple organizational entities, including the government, is a central value. Basically, these two countries have conflicting beliefs about the nature of society.

Hypothesize Needs

The needs columns of Figures 8.5 and 8.6 are based on the foregoing discussion of trade disputes and a review of Figures 3.5, 3.7, 3.10, 4.2, and 6.4. As usual, wording was tailored to be compatible with the context of the analysis.

If one compares entries in the needs columns, the United States is focused almost solely on financial objectives, including winning the game by playing according to long-established rules of the game. Japan, on the other hand, is much more concerned with obligations to others, complying with society, and maintaining traditions.

Another contrast, which was noted in Chapter 3, is the United States penchant for total objectivity, which leads to strong feelings that there are "right" answers to virtually every question. Japan embraces both objectivity and subjectivity, which facilitates compromise in the event that it is necessary.

With both needs and beliefs, an important contrast is quite clear. In the United States, individual freedom and accomplishment are valued. In Japan, community building and serving others are valued. These are not new observations. However, this analysis serves to emphasize the critical nature of these differences.

Evaluate Hypotheses

As indicated in the earlier discussions of disputes involving the environment and economy, the hypothesized beliefs and needs underlying trade disputes have not been evaluated in any formal manner. They primarily reflect a compilation of my personal experiences, leavened by the NBP Model, Template, and Analysis Methodology. Thus, my assertions are perhaps best described as well-informed speculations.

For any real analysis of trade disputes, I would expect that evaluation of hypotheses concerning underlying beliefs and needs would be part of the negotiation process, perhaps part of the joint fact-finding advocated by Raiffa (1982) as well as Susskind and Cruikshank (1987). The methods and tools described in *Design for Success* (Rouse, 1991) might be employed in this evaluation.

Modify Alternatives

Earlier discussions indicated that a key element of consensual negotiation involves broadening the negotiation to include a wider set of concerns that the disputing parties value differentially. From this perspective, it strikes me that the United

ATTRIBUTE	PERCEPTIONS	BELIEFS	NEEDS
Viability	Positive	Exports and jobs will grow and bottom line will meet goals	Needs to recoup investments and make profits
	Negative	Growth will not be achieved and/or costs of change will be too high	Needs for continuity and stability
Acceptability	Positive	Growth will occur in valued industries	Needs to stay in paradigm and play by the rules of the game
	Negative	Decline in valued industries, cultural compromise too great, solutions NIH ("not invented here"), and agreement not fair	Needs for continuity, immutable premises, and having own solution adopted
Validity	Positive	Playing field leveled, innovation will prevail, meeting requirements is sufficient, individual responsibility is key	Needs for stability and control, need to exercise own skills, need to be objective
	Negative	Cannot compete and lack of trust	Needs to deal with uncertainty and risks, need for protection

FIGURE 8.5. NBP Template for the United States.

States has viewed the problem too narrowly. Simply leveling the playing field by removing formal trade barriers is likely to be insufficient.

Put another way, removing formal trade barriers will not level the field. Cultural differences in terms of differing needs and beliefs are primary underlying causes of the issues being debated on the surface. It is unreasonable to expect Japan to change needs and beliefs to satisfy United States desires for growth in exports and jobs.

Since the trade imbalances are overwhelmingly in Japan's favor, the United

ATTRIBUTE	PERCEPTIONS	BELIEFS	NEEDS
Viability	Positive	Exports and jobs will not suffer	Needs to fulfill obligations to others and comply with society
	Negative	Too much will be given up and/or costs of change will be too high	Needs to recoup investments and make profits, needs for continuity and stability
Acceptability	Positive	Trends in valued industries will be positive	Need to fulfill obligations to others
	Negative	Barriers to growth of valued industries, cultural compromises too great, agreement not fair	Needs for continutiy, social consensus, and fitting into tradition
Validity	Positive	Market will drive requirements, value and quality paramount, community should help, long-term solution desired	Need to fulfill obligations to others, needs for stability and control, need to embrace both the objective and subjective
	Negative	Lack of trust, lack of confidence in other parties	Needs to deal with uncertainty and risks

FIGURE 8.6. NBP Template for Japan.

States has to be the one to adapt. In fact, this process is already under way as companies attempt to become market oriented and focus on delighting customers. Discussions in earlier chapters, as well as *Design for Success* (Rouse, 1991) and *Strategies for Innovation* (Rouse, 1992), describe some of the types of change under way.

Another element of adaptation should be reconsideration of the role of government in these areas. To an extent, the 1992 presidential elections in the United

States provided an opportunity to renew this debate. A variety of efforts are now under way to assess alternative approaches to government involvement—I am a member of one of the committees charged with making these assessments.

My guess is that the United States will move more toward the Japanese model of supporting industry. The Japanese model will not be copied, nor will the German model. Instead, lessons learned from these other countries will be incorporated in the design of an American approach that will, in effect, take into account the beliefs and needs underlying United States culture.

Modify Situations

For the longer term, I think that we need to nurture more communitarian values in the United States. As discussed in *Strategies for Innovation* (Rouse, 1992), serving the marketplace should become the vision with financial metrics primarily serving to make sure that we are on the road toward this vision. This change of perspective can greatly enhance opportunities for innovation.

To some extent, emerging environmental problems have helped to foster more community orientation. Interdependencies are becoming increasingly clear. This is also obviously true for economies. We now need to focus on community building and creating as many win–win situations, and as few win–lose situations, as possible.

Plan Life Cycle

Settling trade disputes of the form discussed in this section is not simply a matter of negotiating and signing an agreement, for example, to eliminate tariffs. The process of recognizing and adapting to the evolving rules of a changing game requires a process-oriented perspective. A plan is needed for measuring and refining measures of viability, acceptability, and validity. Similarly, a plan is needed whereby the nature of agreements and forms of adaptation can be modified as measurement results become available. Methods and tools to accomplish these types of planning are discussed in *Design for Success* (Rouse, 1991).

Summary

This section has focused on trade disputes. Sources of trade imbalances between the United States and Japan served as the primary example. In this illustration, we saw that the essence of the dispute was not formal barriers to imports. Instead, deeply ingrained cultural differences appeared to be primary underlying causes. Consequently, the resulting approach to settling such disputes focused on adapting to cultural differences. While this is a "tall order," it seems to be a process that is already under way.

IMPLICATIONS OF ANALYSES

The two case studies discussed in this chapter provide an interesting basis for reflecting upon the underpinnings of the NBP Model. It seems reasonable to conclude that expectations, attributions, and mental models are central to the two types of sociotechnical dispute discussed.

Expectations play an important role in the behaviors that we anticipate of disputing parties. Attributions underlie many of our explanations of the behaviors of disputing parties. In many situations, expectations and attributions can serve to exacerbate rather than mitigate disputes.

Disparities among mental models appear to be prevalent. Models related to the economy, the roles of government, and responsibilities of individuals, groups, organizations, and companies underlie many of our concerns. When disputing parties have incompatible models, it can become difficult to communicate. As a result, it can be virtually impossible to reach a lasting agreement.

Differences concerning the roles of government are often critical. Perspectives range from (1) the government should not be involved from a philosophical point of view, (2) the government cannot be effective even if it should be, (3) the government has to be involved and do the best it can, and (4) the government is the best way to solve a problem in a non-self-interested way. Settling trade disputes, for instance, can be highly influenced by which of these perspectives is adopted. In some case, differences among these perspectives consititue the essence of the dispute.

Ideology also plays a central role in many disputes. In terms of beliefs, it can drive needs. In turn, needs can frame beliefs. Hence, sociotechnical disputes can involve an underlying agenda that goes beyond the concerns on the surface of the dispute. This phenomena is discussed at length in Chapter 9 in the context of political conflicts.

SUMMARY

This chapter has focused on sociotechnical disputes and the use of the NBP Model, Template, and Analysis Methodology for addressing such disputes. It should be clear that much background information is a necessary precursor to performing an NBP analysis. This will be even clearer in Chapter 9.

The role of the NBP analysis is, therefore, to guide information collection and provide a framework for organizing the results of digesting the information obtained. Most of the expertise involved in such an analysis comes from the analyst. Prompting and organization are provided by the methodology. The use of the methodology is discussed in considerable detail in Chapter 10.

REFERENCES

Arrow, K. J. (1963). *Social choice and individual values.* New York: Wiley.

Barke, R. P., and Jenkins-Smith, H. C. (1992). *Politics and scientific expertise: Scientists, risk perception, and nuclear waste policy.* Atlanta, GA: Georgia Institute of Technology, School of Public Policy.

Flynn, J., Kasperson, R., Kunreuther, H., and Slovic, P. (1992). Time to rethink nuclear waste storage. *Issues in Science and Technology*, 42–48.

Hofstede, G. (1980). Motivation, leadership, and organization. Do American theories apply abroad? *Organizational Dynamics*, 42–63.

Moore, C. (1992). Bush's nonsense on jobs and the environment. *New York Times*, September 25, p. A13.

Pate, M. E. (1983). Acceptable decision processes and acceptable risks in public sector regulations. *IEEE Transactions on Systems, Man and Cybernetics, 13*, 113-124.

Peck, M. S. (1987). *The different drum: Community-making and peace.* New York: Simon and Schuster.

Raiffa, H. (1982). *The art and science of negotiation.* Cambridge, MA: Harvard University Press.

Rouse, W. B. (1991). *Design for success: A human-centered approach to designing successful products and systems.* New York: Wiley.

Rouse, W. B. (1992). *Strategies for innovation: Creating successful products, systems and organizations.* New York: Wiley.

Shapira, P. (1992). Lessons from Japan: Helping small manufacturers. *Issues in Science and Technology*, 66–72.

Shapira, P., Roessner, J. D., and Barke, R. (1992). *Federal-state collaboration in industrial modernization.* Atlanta, GA: Georgia Institute of Technology, School of Public Policy.

Smothers, R. (1992). Plan to destroy toxic weapons polarizes a city. *New York Times*, September 24, p. A8.

Susskind, L., and Cruikshank, J. (1987). *Breaking the impasse: Consensual approaches to resolving public disputes.* New York: Harper and Row.

Thurow, L. (1992). *Head to head: The coming economic battle among Japan, Europe, and America.* New York: Morrow.

Chapter 9

Resolving Political Conflicts

This chapter discusses two case studies that are intentionally speculative. These case studies focus on political conflicts in South Africa and the Middle East. I chose these two examples because I have quite a bit of experience in both regions, particularly South Africa, and there is a wealth of reference material available on both of these conflicts.

These examples of our fourth class of archetypical innovation problems, that is, resolving political conflicts, are especially rich in opportunities to apply the Needs–Beliefs–Perceptions (NBP) Model, Template, and Analysis Methodology. Conflicting belief systems are readily apparent on the surface, and subtly identifiable below the surface. Religion, culture, politics, and economics play roles in these conflicts.

For other types of political conflicts, e.g., across organizational and professional cultures, different factors may play roles. For instance, professional disputes are likely to be influenced strongly by disciplinary belief systems, as illustrated in Chapter 4. Discussion in this chapter, however, is limited to political conflicts at national levels.

One way to consider these types of political conflicts is to reference Howard Raiffa's (1982) compilation of attributes of conflict situations (see Fig. 8.1). There are roughly two parties in these conflicts. However, as later discussion shows, these parties are far from monolithic. Thus, as Raiffa notes, the bilateral negotiations of concern involve three agreements—one across the table and one on each side of the table as each party attempts to reach consensus.

There are linkage effects among a variety of issues to be negotiated. While the voting franchise is the predominant issue in South Africa, and land is perhaps the major issue in the Middle East, these issues are central precisely because they interact with so many other issues of concern.

An agreement is desired, which will hopefully be ratified. Threats are possible and frequent. For many of the blacks in South Africa and the Palestinians in the Middle East, there are time-related costs of not improving their situations. On the other hand, the whites in South Africa and the Israelis in the Middle East might like negotiations to proceed as slowly as possible.

Contracts will be binding, but perhaps difficult to enforce. Negotiations have been, and will continue to be, both private and public. Group norms have, to an extent, been agreed upon. However, the lack of monolithic groups makes agreeing to norms, and holding to them, quite difficult.

Finally, third-party intervention is not only possible, but probably required if progress is to be made. The possible roles of third parties—facilitation, mediation, and arbitration—were discussed in Chapter 8. The nature of third-party intervention in each of these two conflicts is discussed as each case study is elaborated.

Therefore, our two examples fit into Raiffa's (1982) framework, which serves to sensitize us to a variety of issues. For instance, the lack of monolithic groups and the plethora of relevant issues make it clear that third-party help is needed. Also, we see that the parties on each side of these disputes are likely to have very different senses of urgency.

The overall strategy advocated by Raiffa (1982), as well as Susskind and Cruikshank (1987), is to attempt to transform win–lose situations into win–win situations. Both Chapters 1 and 8 emphasize the importance of this approach. The process of transforming the situation to win–win can be facilitated by assessing each party's "package" in terms of perceptions of viability, acceptability, and validity.

The necessary transformation is more likely if the result provides more benefits or benefits not possible with the status quo. Also of significance may be defining "win" relative to the likely future without resolution, rather than relative to the status quo. Hence, a central aspect of the process of negotiation, at least for these two examples, is creation of the shared understanding that the current status quo is not one of the options.

The time horizon considered in the negotiations is crucial in two ways. One concern is the initiation and schedule for change. In both of these disputes, there is one party whose constituency wants change now. In contrast, the opposing party in each of these two case studies is likely to want to proceed slowly, since the status quo is, for the most part, in their favor.

Another aspect of the time horizon relates to the profile of future benefits and costs associated with an agreement. A large discrepancy between the time horizons of disputing parties can be both a problem and an opportunity. It is a problem to the extent that the party with a short-term orientation will not make any long-term commitments, or perhaps is unlikely to honor long-term commitments. On the other hand, such a discrepancy can be an opportunity if one party is willing to forego short-term benefits for long-term gain, while the other party values the opposing trade-off. Such differences can form the basis for creative packaging of agreements.

A theme that emerges repeatedly in this chapter concerns the difference between what is perceived to be the basis of a conflict and what in fact underlies a conflict. A crucial element of creating win–win packages involves getting below the surface of perceptions. This may require determining the economic, social, and cultural disparities that underlie what is *apparently* a religious dispute. The fundamental problem may not be theological, but instead the result of the phenomenon discussed in Chapter 3 and characterized this way by Neibuhr (1929): "Castes make outcasts and outcasts make castes." This tendency is likely to occur in combination with another phenomenon noted in Chapter 3, namely what Joseph Campbell (1988) refers to as interpreting myths and models as facts. These two common tendencies can "muddy the waters" relative to understanding the underlying basis of conflicts.

James Laue (1992), at a recent conference organized by the Institute for Peace, suggested 10 principles for conflict resolution which are summarized in Figure 9.1. First, he notes that conflict is universal. We should not expect the world to be free of conflict. However, we should attempt to deal with conflict through some form of negotiation or joint problem solving. Such negotiations should be understood as part of the political process.

Many forms of negotiation can be successful. A key to success is creating agreement packages that satisfy the *underlying* (my emphasis) needs and interests of the conflicting parties. Success also requires that the roles of the various parties to the negotiation be understood. Different roles require representing different points of view, which may or may not be the points of view of the particular people filling these roles.

Resolution of conflicts requires a process that evolves over time and involves different responses at each stage. In this process, it is important that the appropriate "unit of analysis" be identified—otherwise, apparently external parties may undermine the process. The focus of the process is creative resolution rather than winning. Creating a forum where this kind of problem solving can happen tends to be the most critical challenge.

Laue's (1992) principles set the stage for the remainder of this chapter. As noted earlier, the focus is on two long-standing political conflicts—South Africa and the Middle East. With each of these case studies, the analysis begins by summarizing the long histories leading up to the current conflicts. Possible agreements are then formulated using the NBP Model, Template, and Analysis Methodology. The results provide a variety of interesting insights into the nature of these conflicts and how resolution might be approached. This chapter concludes

- Conflict of some kind is universal in all human systems.
- True and full resolution of a conflict occurs only through negotiation or some other form of joint problem solving involving all parties to the conflict.
- Negotiation itself is a political act set within a series of ongoing political relations among parties.
- Many types of negotiation problem solving and processes can work.
- Full resolution, by joint agreement, satisfies the underlying needs and interests of the conflicting parties and does not sacrifice any of their genuinely important values.
- Understanding and respecting the differences between the roles of mediator, advocate, and enforcer are important steps toward real resolution.
- Conflicts proceed in sequences or stages and different responses are appropriate in different stages.
- It is important to identify the appropriate "unit of analysis" in understanding a given conflict, e.g., nation-states, religious or tribal groups, ethnic aggregates.
- Conflict resolution, as its name suggests, is geared toward resolution rather than winning, and its strategic focuses are on analysis, problem solving, and negotiation.
- Creating a forum conducive to negotiated problem solving is the most critical, difficult task in working to resolve deep-rooted conflicts.

FIGURE 9.1. Principles of conflict resolution (adapted from Laue, 1992).

by reconsidering the basis of the NBP formulation, in light of the results of these analyses.

SOUTH AFRICA

In South Africa, the white population, which is very much a minority, has been in power for hundreds of years. The blacks, who comprise the vast majority of the population, are not allowed to vote and, hence, have no direct political power. The coloreds —those of mixed race—and the Indians have only token political power.

The conflict of interest in this section concerns the transition, probably in 1994, to majority rule. This conflict is ostensibly about political enfranchisement. However, there are many more central issues. Economics is central. The standard of living of most whites is that of a "first world" developed country. For most nonwhites, the standard of living is that of a "third-world" developing country. Thus, the conflict is also a "haves" vs. "have-nots" dispute.

Resolution of the conflict is also complicated by the nation's 350-year history. Debate often dissolves into assertions about who was in South Africa first and when. The effects of colonial rule, by the Dutch and the English, for almost 300 of these years also have an overwhelming influence. Until recently, ethnic rivalries, such as between the Xhosa and Zulu, were suppressed and consequently not resolved. There is a feeling among long-oppressed groups that they are owed some form of reparations by their oppressors.

A stable and long-term resolution of this conflict requires that Laue's (1992) principles (Fig. 9.1) be considered. In the long term, all of the parties involved, including the whites, will be much worse off if a mutually satisfactory agreement is not reached in the next few years. This section explores perceptions of alternative agreements. The goal, as noted earlier, is to examine the nature of these perceptions in terms of needs and beliefs.

It is essential that we begin by consideration of the history of South Africa and how the conflict emerged. The material that follows was drawn from a variety of general sources, as well as Martin Meredith's *In the Name of Apartheid* (1988), James Michener's *The Covenant* (1980), and a number of the novels of Wilbur Smith. The fictional depictions of South African history were invaluable for gaining an appreciation of the perspectives of the participants in the evolution of the country's history.

We start in 1652, although, obviously, the history of the region began long before this date. The Dutch East India Company needed a supply point on the long ocean voyage between Holland and Java. Consequently, it stationed Jan van Riebeeck at the Cape of Good Hope, where he was in charge of reprovisioning the company's ships.

At that time, the Khoikhoin tribe occupied most of coastal South Africa. The Dutch bartered with them for sheep and cattle. The Khoikhoin regularly stole the animals back. In this way, a long history of disputes with native Africans began, and intensified as migration on both sides led to clashes.

In 1685, 200 Huguenots arrived, fleeing religious persecution in France. They were relatively quickly assimilated by the Dutch. By 1707, there were roughly 2,000 whites and 1,000 imported slaves in the colony. At this point, the Dutch stopped encouraging immigration to South Africa.

By 1795, the populations had grown to 15,000 whites, who regarded themselves as South African and Afrikaans (a variant of Dutch) as their mother tongue. The company provided the Afrikaner farmers (Boers) with 6,000 acres for a small annual fee. However, due to the poor nature of the land, the Boers had to move every few years. This led to continual expansion of white settlements, and continual battles with native Africans as their lands were taken.

A primary result was a nomadic group of a few thousand farmers spread over 100,000 square miles. The Boers were extreme individualists and strict Calvinists. Their Calvinism was basically a 1650s version because, to a great extent, they had been isolated from the evolution of Western society in general. This isolation continued for at least another 100 years.

In 1795, the British took over the Cape to keep the French out during the Napoleonic Wars. The British left in 1802 when these wars ended. As the Dutch East India Company was now defunct, the Dutch government now ruled the colony. In 1806, the British again took over the Cape.

Over the next 20 years or so, the British introduced a variety of laws to protect slaves. There were trials of Afrikaners for cruelty. The eventual clash between the English and Afrikaners was set in motion.

In 1820, roughly 5,000 British settlers arrived at the Cape. These were the first whites not assimilated by the Afrikaners. At this point, over 100 years had passed without an infusion of "new blood." It can easily be argued that this isolation continued until at least the Anglo-Boer War at the turn of the century.

The British emancipated all slaves in 1833, 30 years before the United States did so. The Afrikaners chafed under the new liberal laws. They wanted to preserve a "doctrine of purity" based on their Calvinistic interpretations of the Old Testament.

In the period 1835–1842, approximately 12,000 "Voortrekkers" left the Cape area and moved north and east. After many bloody battles with the Zulu and Ndebele, they settled in the Transvaal (high veld) to the north and Natal to the east. When the British annexed Natal in 1842, the Voortrekkers retreated to the high veld.

By the early 1850s, the British recognized the independence of the Voortrekkers in Transvaal as well as Orange Free State to the south. By the late 1850s, there were 20,000 Voortrekkers north of the Vaal River. The South African Republic was formed under Andries Pretorious. The constitution of the new republic permitted no racial equality in church or state.

In 1867 diamonds were discovered; in 1886, gold. By 1891, the diamonds were all controlled by DeBeers and the gold was all controlled by a few corporations. Cecil Rhodes was a very central player in this consolidation. His resulting wealth provided the capital for supporting Rhodes scholars at Oxford University in England.

By the late 1890s, Britain wanted to create a grand union of all of southern Africa, urged on in part by Cecil Rhodes. The Afrikaners opposed this push, still reeling from the huge population growth due to the gold rush. They felt that their culture was about to be swallowed by immigrants and British influence.

The resulting frictions precipitated the Anglo-Boer War in 1899. The conflict involved 500,000 British troops and 65,000 Boer troops. British victories quickly led to Boer guerrilla warfare, which the British found to be "improper." Consequently, the British general Kitchener adopted a "scorched earth" strategy. Farms of Boers were completely destroyed. Their families were herded into concentration camps. Over 20,000 women and children died. The war ended in British victory in 1902, but divisions and hatred were now deep-seated due to Kitchener.

In 1910, the four colonies—Cape Province, Natal, Orange Free State, and Transvaal—became the Union of South Africa. At roughly the same time (1911–1914), Gandhi was leading a nonviolent tax revolt, honing his skills for later use in India. (In the United States, Martin Luther King Jr. would subsequently adapt his nonviolent tactics for use in the Civil Rights movement). Britain was, of course, the common link among all these trends and events, exacerbating problems with her imperial tendencies and promoting change with her liberal leanings.

Nineteen twelve was a pivotal year, if only in retrospect. The Nationalist Party (NP) was formed by Afrikaners that year, and the African National Congress (ANC) was formed by blacks. Eighty years later, these parties are the lead players in the unfolding transformation of South Africa.

In the forty years between 1910 and 1948, the Afrikaners consistently opposed the liberal drift. Not surprisingly, they also opposed relations with Britain. In 1948, Daniel Malan and the NP won the elections. Apartheid began. A mass of legislation was enacted to preserve white supremacy, including separate living, education, marriages, transportation, and recreation. The Dutch Reformed Church provided theological legitimization for apartheid with its creative interpretations of the Old Testament. (This position was recanted in 1989—I can easily recall the announcement.) Other denominations protested these initial theological claims and the emergence of apartheid, but not too loudly. As noted in Chapter 3, institutional churches tend to support the status quo, regardless of the theology implied.

In 1951, the coloreds were removed from the voting rolls. Soon after, the Senate was enlarged to give the NP two-thirds of the seats. By 1959, all African participation in the government had been eliminated throughout the country.

Throughout the 1950s, the ANC fostered passive resistance, such as a bus boycott in 1957. In 1960, the ANC and other black parties were outlawed. Nelson Mandela, Walter Sisulu, and seven other black leaders were sentenced to life imprisonment in 1964.

In the 1970s, the black consciousness movement grew with the leadership of Steve Biko. This movement focused on blacks gaining self-identity, rather than aspiring to a white identity. In 1976, there were riots in Soweto, protesting the requirement that all children, regardless of race, learn Afrikaans in school. In 1977, Steve Biko died due to a police beating.

The 1980s saw immense world pressure on the Afrikaner government to reform. In 1984, a tricameral parliament was created with white, colored, and Indian chambers. Blacks were still not enfranchised. In 1985, economic sanctions were imposed—they had originally been approved by the United Nations in 1962. As nation after nation imposed their version of the sanctions, South Africa's economy steadily waned.

In 1989, F. W. de Klerk was elected President. He immediately began a program of reform. Nelson Mandela was freed in 1990 and assumed the leadership of the ANC, which was no longer outlawed. With additional freedom—but still no vote—old tribal rivalries flared with conflicts between the Xhosa and Zulu.

At the present time, South Africa's 30 million blacks, who represent about 85 percent of the population, are anticipating the opportunity to participate in the next elections in 1994. In fact, they are anticipating opportunity in general, as well as solutions to problems such as very high unemployment and an extreme lack of adequate housing. Expectations are quite high, as are tensions. The on-again, off-again negotiations of the NP and the ANC lead to charges, countercharges, and occasional hopeful outcomes.

The current conflict in South Africa is not as straightforward as I have described it thus far. The wide variety of opinions on both sides of the conflict is well illustrated in Alan Fischer and Michel Albeldas' (1987) book of interviews with prominent South Africans. Recent articles by Bill Keller (1992a, 1992b) also reinforce the complexity of the issues involved.

In one article (1992a), he notes that liberal whites (many of them of British origin) are becoming disenchanted with the ANC, which wants majority rule but does not necessarily understand democratic values such as free press, free speech, and free association. The ANC is a liberation movement, not a liberal organization. Keller asserts that the style of the ANC is more charismatic than cerebral. Put simply, he concludes that the ANC's black following was not raised on Jeffersonian democracy and does not inherently value the same things as the liberal whites.

In another recent article (1992b), Keller reports that many colored (mixed-race) voters are now seeing the NP as representing security, morality, and prosperity. The ANC, on the other hand, increasingly represents violence, communism, and economic redistribution. He notes that they fear the rise of the blacks more than oppression of whites.

Thus, each of the two parties in this conflict is by no means monolithic. Using Raiffa's (1982) guidelines, this suggests that three agreements are needed—one across the table and one on each side of the table. Transforming this conflict from win–lose to win–win will require ingenuity.

Fortunately, there is some strength to build upon. While the current economic situation in South Africa is bleak, the long-term prospects, according to James Henry (1990), are promising. Eliot Janeway (1990) asserts that South Africa offers much better investment opportunities than, for instance, Eastern Europe. Flora Lewis (1990) discusses the regional role of South Africa, noting that the country can become the economic powerhouse of the sub-Sahara if it embraces democracy and emerges from isolationism.

In the past five years, I have spent quite a bit of time in South Africa. After my first few visits, I was asked to address a group of South African business executives (Rouse, 1989). They were interested in the perspective of an American businessman. They were also aware of my visits in the black community and wanted to know what I had learned during those visits.

In my presentation, I used the theme "black and white" on three levels. First, I explained Americans' penchant for wanting to see all issues as simple, crisp decisions. Second, I suggested that the black and white of the media (e.g., the newspaper) was the filter through which most Americans viewed South Africa. Finally, of course, I focused on the black and white of racial differences.

In this part of my remarks, I suggested that they should start working with the black community immediately, if only with small projects. The results would include whatever good the project yielded, and more importantly, much communication among the project participants. They would no longer have to ask foreign visitors, such as myself, what the black community was thinking.

More recently, after many additional visits, I prepared a working paper that was distributed to a variety of business leaders (Rouse, 1991). I summarized my observations, with particular emphasis on the upcoming 1994 elections. First, I considered the problems on the surface. The most obvious problem is the need to enfranchise all people in South Africa. While this need is important to the currently disenfranchised blacks, there are, to an extent, more pressing problems: jobs, education, and housing. Problems looming in the background include population growth and AIDS.

Below the surface, the aforementioned communication gulf has resulted in blacks and whites having very muddled expectations of each other. As a result, whites are uncertain about blacks' priorities and what blacks will do when they have political power. Blacks, quite understandably, are extremely cautious in trusting whites.

Confidence could be built if white businesses started investing in activities that yield jobs, foster improved education, and provide the means for housing. However, this is a tall order in a weakened economy. Further, many businesses do not have experience in the type of free market that is emerging.

In this paper (Rouse, 1991), I concluded that the key elements needed are job creation via new ventures, entrepreneurial training to assist new ventures, black–white cooperation in forming these ventures, and focus on expediting the whole process. These ingredients, I argued, are the means for tangible progress leading up to the 1994 elections.

This working paper led to a variety of thoughtful responses. All of them were supportive. One of them indicated that the paper was being distributed among a broader set of business leaders, politicians, and civil servants. This, not surprisingly, was my intent.

This discussion of my own experiences is not meant to imply that I have any magic solutions to offer in this complicated conflict. My purpose in relating these experiences is simply to articulate the background upon which the following analysis is based.

Identify Measures

Measures of viability relate to benefits gained and costs incurred. Benefits sought include the basics of food, clothing, shelter, and overall standard of living, which can be translated into jobs and wages. Also of interest are safety and opportunity. Safety relates to physical harm and political oppression. Opportunity is concerned with freedom of movement—being able to live in any locale and work in any vocation.

Costs include the necessity of changing, as well as the time and effort involved. It appears that the costs will inevitably be much greater for whites than blacks. However, if resolution of the conflict results in dramatic economic growth, as well as much greater political stability, the economic pie may increase sufficiently fast to meet black expectations without totally draining whites' resources.

Expectations play a key role in perceptions of viability. Few, if any, people in South Africa expect the status quo to survive. Whites expect some form of loss, at least in the short term. Blacks expect considerable gain, hopefully sooner than later. A creative agreement may enable whites' losses to be less than expected—providing a gain relative to expectations. In contrast, blacks' expectations of gains are likely to be very difficult to satisfy.

Measures of acceptability relate to the extent that valid and viable solutions are also acceptable. Enfranchisement of all citizens of South Africa may be accomplished in a way that provides much greater benefits than costs, but be unacceptable because basic democratic values are not maintained. For example, threats to a free press and free speech would cause liberal whites to find such a plan unacceptable. Also unacceptable to whites would be a one-party state such as resulted in Zambia and Zimbabwe, for instance.

For many blacks, attempts to homogenize all groups would be unacceptable. The black consciousness movement caused many groups to claim their black identities rather than attempt to gain white identities. Therefore, a solution that requires, in effect, all parties to become Afrikaners is unlikely to be acceptable.

Validity concerns the extent to which a plan is viewed as solving the problems of interest. On the surface, the problem is one of providing universal suffrage. Below the surface, however, the problem concerns power. Blacks will perceive a solution to be valid if they gain political power and control of the country. Whites, on the other hand, will question the validity of a solution that may lead to a "tyranny of the majority."

Determine Alternatives

All of the alternatives concern how universal suffrage is implemented. Blacks want "one man, one vote." Whites want two types of assurance. First, they want to avoid the "one man, one vote once" syndrome that occurred in many African countries that gained independence over the past 30 years. In these countries, it was typical for the predominant tribe to gain control, declare a one-party state, and henceforth have no competition for political leadership.

Whites' second, and more important, concern relates to the potential for their complete loss of power. Whites have argued for a system whereby they will have some measure of power despite their being only 15 percent of the population. They want veto power, which seems unlikely. At the very least, whites want a representation scheme that enables predominantly white areas to elect white representatives.

Assess Perceptions

All of the variations that I am aware of provide for universal suffrage. Hence, they are probably perceived to be valid by blacks. On the other hand, whites tend to question the validity of solutions that strip them of all power. A basic problem is whites' lack of confidence that blacks can govern effectively. Thus, they feel that white concerns will not be addressed and that government will slow to an inefficient snail's pace.

With regard to acceptability, the two main problems for blacks are (1) the possibility of whites having veto power and, in this way, overruling any legislation that compromises white interests, and (2) the fact that implementation of plans seems very slow. For whites, the primary concern is whether elections will result in a true democracy, or simply black rule.

Perceptions of viability concern perceived benefits and costs relative to what is likely without a negotiated resolution. While blacks may perceive current plans as providing many benefits, the actual growth of jobs and wages is likely to be much slower than hoped. Accordingly, blacks' perceptions of viability are likely to move from positive to negative as implementation proceeds. While many whites are currently pessimistic, it is quite likely that benefits and costs will be more attractive than anticipated as implementation unfolds.

Define NBP Templates

For this analysis, two NBP templates are used—one for blacks (Fig. 9.2) and one for whites (Fig. 9.3). For a real application of this methodology to the South African conflict, there might be several templates. On the black side, there might be one for the ANC and one for Inkatha (the Zulu party), as well as one for coloreds and one for Indians. The white side might include templates for conservative Afrikaners, liberal Afrikaners, and the Anglos.

ATTRIBUTE	PERCEPTIONS	BELIEFS	NEEDS
Viability	Positive	Jobs and wages will grow substantially with change, education and housing will improve substantially	Need for basic food, clothing and shelter, need to socialize young
	Negative	Jobs and wages will not grow substantially with change, education and housing will not improve substantially	Same as above
Acceptability	Positive	Ethnic identity will be enhanced	Need to belong, fit into tradition, and be secure in membership, need for social redemption
	Negative	Will remain outcast, ethnic identiy will be undermined, process will take too long	Same as above
Validity	Positive	Enfranchisement will lead to freedom, self-identity will be established, process will be fair	Need for power and to be in control, need to communicate and participate
	Negative	One man, one vote will not lead to freedom, white veto power will not allow freedom	Same as above

FIGURE 9.2. NBP Template for blacks in South Africa.

ATTRIBUTE	PERCEPTIONS	BELIEFS	NEEDS
Viability	Positive	Standard of living will be greater than anticipated, i.e., losses will be smaller	Need for continuity, needs to recoup investments and make profits
	Negative	Standard of living will decrease substantially and status will be diminished	Same as above
Acceptability	Positive	Afrikaner culture will be preserved and "chosen" people will prosper	Need for continuity and stability, need to belong and be right
	Negative	Single party will result without democratic values, change will occur too quickly	Need to be in control, needs to avoid risks, uncertainty, and unknowns
Validity	Positive	Process will be fair and whites will retain some power in democracy	Need for protection, need to communicate, need for power
	Negative	Tyranny of majority will result, black government will be incompetent	Need for power, needs to avoid risks, uncertainty, and unknowns

FIGURE 9.3. NBP Template for whites in South Africa.

Hypothesize Beliefs

As usual, entries in the beliefs columns of Figures 9.2 and 9.3 were determined by reviewing the discussion of the problem thus far, and by consulting Figures 3.6, 3.11, 4.3, and 6.5. There is an important contrast between the beliefs columns of Figures 9.2 and 9.3. Blacks are seeking freedom and access to jobs and wages. Whites are seeking a more liberal democracy and preservation of their standards of living. While these goals are not incompatible, they also are not identical.

Hypothesize Needs

The needs columns of Figures 9.2 and 9.3 were derived from reviewing the discussion of the problem and consulting Figures 3.5, 3.7, 3.10, 4.2, and 6.4. Notice that in a few cases I have indicated "same as above" because similar needs can underlie both positive and negative perceptions. The "art" of making such determinations is discussed in Chapter 10.

Evaluate Hypotheses

Other than personal experiences, I have no data with which to evaluate the veracity of the needs and beliefs hypothesized. In an actual negotiation process, these hypotheses could be tested using questionnaires and interviews. Role playing might also be a way to illustrate the impact of needs and beliefs on perceptions, as well as solicit reactions to the needs, beliefs, and perceptions depicted.

Modify Alternatives

In order to transform this win–lose situation into a win–win situation, it is probably necessary to broaden the negotiation to include issues other than enfranchisement. The white business community, which potentially has the ability to provide the jobs and wages sought by blacks, has had difficulty obtaining investment capital. They also have entertained thoughts of moving their assets out of South Africa as a hedge against possible nationalization.

If the negotiation can be broadened to include business considerations, from both national and international perspectives, it is possible to make change attractive to all of the parties. Blacks will have to commit to multiparty democratic values, as well as policies that do not affect the private sector negatively. At the same time, whites will have to commit to job creation and wage growth as rapidly as possible. Tax rates might be linked to such metrics, that is, taxes would decrease as the number of jobs and wages increased.

The international business community could help this process in two ways. First, it could provide investment capital which, as noted earlier, should be attractive once political stability is achieved. Second, continued investment could be linked to progress, after a reasonable period of time, in productivity, quality, and creation of skilled workers.

Government and industry could facilitate this process by making major commitments to education and training. The generation that spent its youth and young adulthood protesting rather than in school sorely needs education and training. The focus should be on skills that support gaining jobs in industries where wage growth is likely.

This proposal will require much cooperation among parties that have a long history of distrusting one another. Why should they now commit to such cooperation? There are two reasons. First, by employing the types of negotiation processes advocated by Raiffa (1982), and Susskind and Cruikshank (1987), it should be possible to convince them that cooperation will lead to a win–win rather than a win–lose outcome.

Second, by expanding the negotiation beyond South Africa's borders, more assets and leverage can be brought to bear. South Africans have been isolated for many years. In fact, they have a tradition of isolation, as noted in earlier discussions. The outside world, via investment capital and technical assistance, can make it attractive for both sides of the conflict to cooperate, for the mutual gain of everyone involved.

Modify Situations

For the longer term, it should be possible to modify parties' attitudes toward one another. If jobs, housing, and particularly education can be dramatically improved by cooperation, then basic needs will have been satisfied. Moreover, questions of identity and preservation of culture will work themselves out.

My guess is that it will take 10 years from the 1994 elections for the situation to have changed substantially. But, the first glimmers of the needed changes could happen in one or two years. If these initial successes are orchestrated appropriately, with much credit given to all of the parties involved, the vast majority of people whose expectations will not initially be met are likely to perceive the promise as delayed, but not broken.

Plan Life Cycle

It will be difficult for the newly elected government to plan in detail the types of changes that I have described. Much will depend on free or semifree markets playing their roles. However, the government can lead this process. What is needed is a well-articulated vision of what South Africa will be like in 5, 10, or 20 years. This vision should be compelling, but realistic. It should also portray people's roles and activities in those future times.

This story might be written in the iterative manner that I described in Chapter 7. Specifically, different constituencies might be asked to react to evolving versions of the vision. They could ask questions, make suggestions, and criticize in general. After many months and, surely, many iterations, a consensus vision is likely to emerge. This vision would be "owned" by hundreds or perhaps thousands of people from all sectors of society. These same people would then, in their own ways, contribute to making this vision a reality.

Admittedly, this sounds farfetched. Who would do the writing of such a story? What language would it be in? These are straightforward but difficult questions. I cannot answer them. However, I promised that this chapter would be speculative!

Summary

This section has considered the conflict in South Africa as a way of illustrating use of the NBP Model, Template, and Analysis Methodology for addressing political conflicts. While this analysis was necessarily cursory, it did show how beliefs underlying perceptions, and needs related to beliefs, can provide insights into sources of disagreements surrounding the resolution of conflicts. This discussion of South Africa has served as a "warmup" for consideration of what must be the longest-running conflict in the history of humankind.

THE MIDDLE EAST

Conflict in the Middle East has a long history. The conflict involves an interdependent set of territorial, economic, and political issues, interwoven with cultural and religious issues. For roughly 4,000 years, this conflict has hinged on basic questions such as, "Who's first?" and "Who's chosen?" All parties to the conflict feel they know the answers to these questions.

Ironically, all of the parties involved have similar and closely linked histories. Their religions are all monotheistic. Many of the same prophets are revered. The parties also share the same "sense of place," at least in terms of the Holy Land. However, as Joseph Campbell (1988) points out, metaphorical differences obscure common needs and beliefs.

A variety of practical issues are also relevant. Western forms of prosperity and oil-rich opulence contrast with masses of refugees. The give and take of democracy contrasts with dictates of aristocratic or military rule. Basic needs of safety and security are central to everyday life.

There is also a sense that 4,000 years of conflict will inevitably lead to continued conflict. On one of my visits to Jerusalem, an archaeologist recruited by a colleague showed me various ruins and stages of history as displayed by the types of stones and construction methods underlying existing buildings. I commented on the fact that so many conquerors and rulers had come and gone in the city's long history. He said that the "teeth of time" inevitably led to their passing. I was struck by this phrase. Did it mean that positive change was possible—a good omen—or that change was impossible, with one conflict following another? This question is explored, but not fully answered, in this section.

To begin our exploration of this conflict, it is important to review the history of the region. The overview that follows was gleaned from a variety of general sources, including Jimmy Carter's *The Blood of Abraham* (1985) and Thomas Friedman's *From Beirut to Jerusalem* (1989), as well as James Michener's *The Source* (1965). Michener's fictional treatment of Middle East history was invaluable for imagining what it was like to be a participant in the rich history of the region.

As the name of his book implies, Jimmy Carter begins his discussion of the history of the Middle East conflict with the story of Abraham in Canaan in roughly 1900 B.C. According to tradition, Abraham fathered Ishmael by Hagar, Isaac by Sarah, and six sons by Keturah. The line of Ishmael led to the Arabs. The line of Isaac led to the Jews. The other six sons produced lines that led to the tribes in Egypt, Jordan, Lebanon, Syria, and other North African tribes. From this perspective, the Middle East conflict is inherently a family dispute!

Isaac's sons were Jacob and Esau. Jacob became the leader of the family and his name was changed to Israel. The family of Israel moved to Egypt in a time of severe drought (c. 1600 B.C.). After more than 400 years of slavery in Egypt, Moses led the Israelites out (c. 1200 B.C.), and eventually they made it back to Canaan. By roughly 1000 B.C., the 12 tribes of Israel were united and in control of Canaan under King David. Within a few generations, the nation divided into two weaker kingdoms—Israel which included 10 tribes in the north, and Judah, which included 2 tribes in the south. Israel was destroyed by the Assyrians in roughly 700 B.C. Judah was destroyed by the Babylonians in roughly 600 B.C.. In this way, the Jewish dispersion, or Diaspora, began as the Jews were carried off by their conquerors.

It is interesting to note that all three faiths in the region—Judaism, Christianity, and Islam—share much of this history. We do not know the story of Ishmael's line like we do of Jacob's. However, the Jews and Arabs shared the same lands during many of these periods of time. In addition, all three religions revere many of the same stories and prophets.

The Greeks under Alexander conquered the region in 332 B.C. The Jews revolted and formed Judea in 176 B.C. The Romans took Jerusalem and control of the region in 63 B.C. A Jewish revolt in A.D. 70 was put down by the Romans. Further Jewish revolts in A.D. 135 led the Romans to lay waste to Judea. The Romans renamed the region Palaestrina, a name probably derived from the Philistine tribe. The Roman emperor Constantine embraced Christianity in A.D. 313 and, as noted in Chapter 3, the evolution of this sect to a church was accelerated.

By A.D. 650, Muslims controlled the region. The Crusaders took Jerusalem in 1099. The Muslims retook Jerusalem in 1187 and, except for a brief period in the 13th century, controlled the region until the end of World War I. The Ottomans took control in 1516 and continued Muslim rule.

From this terse chronology, it is clear that the region has always been affected by outside influences. Struggles for commercial gain or political benefits have been common. Thus, to a great extent, the "Middle East conflict" is a phrase that has had meaning for several thousand years.

Manifestations of Zionism emerged in the 1880s with settlements in Palestine. Zionism is the movement to reestablish a Jewish nation in Palestine. It draws its name from Mt. Zion, a hill in Jerusalem where the City of David was located. The first Zionist World Congress was held in 1897.

In the midst of World War I, when the Ottoman Empire still controlled the region, Britain issued the Balfour Declaration, promising a Jewish national home in Palestine while also preserving the civil and religious rights of non-Jewish citizens. When World War I ended in the defeat of the Ottoman Empire and control of the region by Britain, the British had to follow through on this promise. The Zionists objected to any limitations on Jewish immigration and land purchases. The Arabs in Palestine opposed the whole undertaking. Over 80 years later, the same debate continues.

From the late 1930s through the 1940s, Egypt, Syria, Lebanon, and Jordan gained their independence. Jewish terrorist attacks on the British, as well as worldwide pressure to provide a home for Jews in the wake of the Holocaust, prompted a request for the United Nations to resolve the issue. A 1947 United Nations resolution, backed by the United States and Soviet Union, resulted in Palestine being divided into Jewish, Palestinian, and international areas, the latter being Jerusalem. In 1948, British control of Palestine ended and the State of Israel was proclaimed.

Soon after, Israel was attacked by its Arab neighbors. The armistice in 1949 gave Israel more land, but the Arabs retained Old Jerusalem. Jordan annexed the West Bank of the Jordan River. Egypt occupied the Gaza Strip along the Mediterranean Sea. In 1956, Israel invaded the Gaza Strip, as well as Sinai to the south. The Israelis were forced to withdraw in 1957 by the United Nations.

In 1964, the Palestine Liberation Organization (PLO) was formed. The express goal of the PLO was to destroy Israel and gain control of Palestine. The PLO has engineered persistent terrorist attacks. Perhaps the most notable of these was the massacre of Israeli athletes at the 1972 Olympics in Munich.

In 1967, Israel reacted to border skirmishes by launching preemptive attacks on Egypt, Syria, Iraq, and Jordan. The Six Day War resulted in Israel occupying the Golan Heights, the Gaza Strip, Sinai, the West Bank, and Jerusalem. The United Nations passed Resolution 242, which required Israel to withdraw from all occupied territories and recognized the rights of all states in the region to exist. Israel rejected withdrawal. The Arabs rejected recognition of Israel.

Civil war erupted in Jordan in 1970, pitting Jordanians against Palestinians. In 1971, the Palestinians were ejected from Jordan and moved to Lebanon. Civil war erupted in Lebanon in 1976 and Syria intervened. In reaction to Palestinian raids on Israel's northern borders, Israel invaded Lebanon in 1982. Soon after, American and European peacekeeping forces entered Lebanon. A terrorist attack on the United States embassy in Beirut in 1983 resulted in 50 deaths. An attack on the

United States Marine barracks lead to over 300 deaths. Lebanon's president Bashir Gemayel was assassinated in 1982. The PLO was ejected from Lebanon in late 1983. Israel initiated withdrawal from Lebanon in 1985.

In late 1973, Egypt attacked Israel in the Sinai, and Syria attacked in the Golan Heights. Both attacks were repelled. The Arabs embargoed oil shipments to the United States—prices quadrupled. U.N. Resolution 338 passed, confirming Resolution 242 and calling for a peace conference. In early 1974, the first Sinai disengagement agreement was signed, placing U.N. forces between those of Egypt and Israel. The second Sinai agreement was signed in 1975. The Camp David agreement was signed in 1978 by Menachem Begin of Israel and Anwar Sadat of Egypt. Sadat's willingness to negotiate with Israel resulted in his assassination in 1981. The Sinai was returned to Egypt in 1982.

A variety of other events occurred in the Middle East during the same period. In 1979, the Shah fled Iran and Ayatollah Khomeini returned, which resulted in Iran becoming a fundamentalist Muslim state. Also in 1979, Soviet troops invaded Afghanistan. The war between Iran and Iraq began in 1980 and lasted until a cease-fire in 1988. Israeli bombers destroyed an Iraqi nuclear reactor in 1981. Iraq invaded Kuwait in 1990, and was expelled in 1991 by forces of the United States and many allies.

This terse history of the Middle East—see the aforementioned books by Carter (1985), Friedman (1989), and Michener (1965) for much more complete stories—illustrates the long-standing and complex nature of the conflict. Neither side of the conflict is monolithic. Moreover, conflict has been a way of life for generation after generation. As Jimmy Carter (1985) states, "The contending parties believe in the rightness of their cause, and some of them are willing to face death rather than change their position or even admit the legal existence of their adversaries. They act with absolute certainty that they are carrying out the will of God." Thomas Friedman (1989) notes, "As long as any party to the Arab-Israeli conflict is focused entirely on obtaining his historical or God-given 'rights,' as he sees them, he is not going to be able to make decisions exclusively on the basis of interests."

Friedman (1989) also asserts that the underlying conflicts are tribelike because "most people have not broken from their primordial identities." Carter notes that "loyalty to family and religious group transcends any commitment to national unity." Friedman claims that people are "hemmed in by ancient tribal and religious boundaries." The resulting tribal logic disdains compromise. Parties to the conflict are not taken seriously unless, as Friedman says, they "are ready to break a little furniture" for their ideas. The result is "a permanent struggle for survival against a hostile world."

In the midst of this struggle, the concept of a Jewish state has been kept alive since the time of King David. During my visits, people have told me that the state, rather than the religion of Judaism, provides Jewish identity for the vast majority of Jews. Consequently, Friedman (1989) notes that "the country has become a living museum of Jewish history."

To a great extent, the West finds this museum to be compelling. Friedman (1989) asserts that this is due to the fact that the world "is always filtered through certain cultural and historical lenses before being painted on our minds." These lenses are "superstories" such as religious and political ideologies. The Bible is the oldest and best-known superstory of Western civilization. Friedman (1989) notes that "the Palestinians simply are not part of the biblical super story through which the West looks at the world."

From the Arab point of view in general, and Palestinian perspective in particular, Carter (1985) notes that strong ideas of nationhood only emerged in the past 50 years, especially as Zionist immigration to the region increased. Many of the Arabs' problems—not just those with Israel—resulted from the arbitrary partitioning of the region by Britain and France, without regard to natural boundaries, ethnic identities, or tribal unity. Consequently, Carter claims, "Despite the common language, customs, and religion, and regardless of the desire of influential and prosperous leaders for harmony and unity of purpose, the Islamic world is still torn by strife that is not limited to combat with Israel." Friedman (1989) asserts, "Islamic fundamentalism and Arab nationalism are the cheap currency, in fact, with which Middle East regimes purchase the lives of the young men."

Nevertheless, the Arab world claims it has ancient rights just as Israel claims it has. Therefore, the Arabs assert, the Palestinians should have same rights to self-determination that Israel has. Yasir Arafat, as head of the PLO, has crystallized this vision for the Palestinians. They, consequently, expect civil and political rights that they are denied on the West Bank. Israel, not surprisingly, argues that such limitations are inherent to military occupation.

Carter (1985) claims that the essence of the problem is that both Jews and Palestinian Arabs want "no less than recognition, acceptance, independence, sovereignty, and territorial identity." He notes, however, that "the situation in the Middle East continues to be unstable because of two crucial factors. First, the Arabs refuse to give clear and official recognition to the right of Israel to exist in peace within clearly defined and secure borders. Second, the Israelis refuse to withdraw from the occupied territories and to grant the Palestinians their basic human rights, including self-determination."

Friedman (1989) captures the Israeli perspective in noting, "[Peoples'] whole lives have left [them] with the conviction that the Arabs would never willingly accept a Jewish state in their midst and that any concessions to the Palestinians would eventually be used to liquidate the Jewish state." He suggests that Israeli beliefs that the primary Palestinian goal is destruction of Israel, rather than building of a Palestinian state, will not allow Israelis to recognize Palestinian rights. Carter (1985) portrays the disputants' views of each other: "Despite—or perhaps because of—[their] vivid similarities of ancient and recent history, Israelis and

Palestinians generally scorn and despise each other and usually deny that there is any parallel between their circumstances. It is as though to recognize in any way the legitimacy of their adversary's case would mean weakening their own."

Where does this leave us? Jimmy Carter (1985) and Herbert Kelman (1987) suggest principles upon which negotiations should proceed. Carter asserts that negotiations must be premised on preserving the security of Israel, recognizing the sovereignty of nations and preserving their borders, resolving the conflict via peaceful means free from terrorism, involving all parties to the dispute, and protecting human rights.

Kelman's principles—which he terms political-psychological assumptions—are much more specific. He asserts that nationhood for both the Jews and Palestinians must be a given, as must be acceptance of each party's diaspora. Sharing the land also must be a given. The leadership on both sides must be accepted as legitimate. Also of central importance for each party is recognition of the distinction between the opposing party's ideological dreams and operational programs. This requires that each party differentiate between negative and positive components of the other's ideology and symbols of legitimacy, which should foster the perception that the sole goal is not to destroy each other.

The remainder of this section focuses on an NBP analysis of the Middle East conflict. Principles such as those suggested by Carter and Kelman are revisited in this analysis, as are a variety of notions discussed in this introductory material. This analysis leads to reasonable conclusions. However, it must be recognized that the type of analytical formulation presented here is, by no means, the normal way that this conflict has been framed.

This point became quite clear to me during one of my recent trips to Jerusalem. In observing many orthodox/fundamentalist Jews, Christians, and Muslims, I found the emphasis on concrete, physical symbols of places and traditions to be almost overwhelming. For example, I saw old women from Cyprus who, my guide said, had likely saved pennies all their lives to visit the Holy Sepulcher. I wondered at their apparent literal acceptance and total commitment to events and places that are of debatable veracity and traditions that, in some ways, are impractical, inconvenient, and anachronistic.

However, in light of the material discussed in Chapter 3, the needs and beliefs of the people I observed seem quite natural. Nonetheless, these observations are very important. The clear implication is that my abstract, analytical approach to this conflict would not be meaningful to almost all of the parties in the conflict. They could not possibly accept their realities as metaphors and, accordingly, would not accept that the issues that dominate their debates are, for the most part, surface features.

And, the fact is, they are right! The analysis that follows provides insights that can serve as catalysts for change. These insights, though, are not the essence of the changes needed. The primary goal is not to change needs and beliefs. The primary value of these insights "below the surface" is to facilitate the changes "on the surface" that the various parties to this conflict seek.

Identify Measures

Often there are several ways that issues can be partitioned into the categories of viability, acceptability, and validity. I choose to limit viability to economic concerns in this case study. Thus, viability can be related to housing, jobs, and wages, which are of particular interest to Israel with its many Russian immigrants. The Palestinians have similar interests and, in some cases, concerns for basic food, clothing, and shelter. Israel is also concerned about possible economic costs of resolving the conflict, especially if investments in occupied areas are forfeited.

Acceptability concerns are mainly related to creation and maintenance of ethnic and national identities. Both sides are concerned that their respective dispersions end. Palestinians are concerned that their leadership be given legitimacy. Economically viable resolutions will not necessarily be acceptable.

I choose to characterize validity in terms of land and sovereignty. Israel wants land and sovereignty to be secure. It also wants civil and political rights to be protected. Palestinians want to gain land and sovereignty, while also gaining civil and political rights.

Notice that I have indicated land to be a validity issue, rather than a viability issue in terms of a benefit. This choice was straightforward. As noted by Kelman (1987), a potential solution will not be considered at all if it does not involve sharing the land. Thus, land is not part of a cost-benefit trade-off—it is a necessity.

Determine Alternatives

The most frequently debated alternative involves the Arabs recognizing and accepting the existence of Israel, and the Israelis withdrawing from occupied territories. Another possibility would allow Israeli settlers to remain in the West Bank, for example, but cede political control to the Palestinians. Yet another possibility is some arrangement involving Jordan and the West Bank.

All of the alternatives hinge on the issue of how agreements will be enforced. How can Israel be sure that the Palestinians or other Arab countries will not attack from a stronger base once Palestine is formed? How can Palestine be sure that Israel will not once again occupy the West Bank and other territories?

There have to be strong disincentives for such actions. One type of disincentive is the use of force, perhaps by the United Nations, to rectify any aberrant behaviors. Economic sanctions are another means. Hence, breaching the agreement could result in a substantial "downside."

Personally, I am more in favor of situations in which loss of "upside" is the penalty. The problem then becomes one of casting the agreement in terms of benefits gained rather than problems solved. The formulation of the analysis has emphasized economic benefits and my guess is that such a focus might be helpful.

Could the Middle East become an economic powerhouse, perhaps a small EEC (European Economic Community)? Certainly, the capital is there among the oil-producing countries. The population is sizable, although its purchasing power is not uniformly strong. While Israel does not have oil or a very large population, it does have substantial human capital, as described in the next paragraph.

Bernard Avishavi (1991) discusses Israel's economic future. He notes that Israel's position has been weakened by long-standing conflicts, excessive subsidizing of industry, and overdependence on Washington. However, Israel has more scientists and technicians per capita and the highest rates of literacy and math skills of any nation on the planet. He argues that the best opportunity for Israeli high-tech entrepreneurs is to get into the R&D programs of global manufacturers. The most important first step is to get on with the peace process. Today, Avishavi argues, "Israel can have the Whole Land of Israel or it can have a piece of the global economy. It can have another West Bank settlement or a Toshiba technology center. It cannot have both."

It seems to me that creative synergies are quite possible whereby the benefits of Jewish–Arab cooperation are much greater than the value of maintaining or renewing discord. Put simply, I am speculating that a Jewish–Arab partnership could result in a win–win situation in which each side would gain more than it would even if it were the winning side of a win–lose formulation.

Assess Perceptions

Perceptions of all of the above alternatives would be very mixed. With convincing third-party guarantees, positive perceptions would be more prevalent. We must, however, keep in mind that we are dealing with needs and beliefs nurtured over thousands of years. The following analysis explores these needs and beliefs.

Define NBP Templates

As the two sides of this conflict are by no means monolithic, a detailed analysis should employ an NBP template for each major party. This might include templates for Egyptians, Jordanians, Lebanese, Palestinians, and Syrians on the Arab side of the conflict. Templates on the Israeli side might include one each for conservative and liberal Jews, and perhaps one for American Jews since, as Friedman (1989) emphasizes, they play a key role in matters relating to Israel.

For the purposes of illustration, however, only two templates are employed. Figure 9.4 is the NBP Template for Israelis; Figure 9.5 is the template for Palestinians. While use of only two templates very much simplifies a complex web of relationships, this simplification enables succinct illustration of use of the methodology.

ATTRIBUTE	PERCEPTIONS	BELIEFS	NEEDS
Viability	Positive	Will have positive economic impact, result in housing, jobs, and wages for immigrants, and open Arab markets	Needs to provide for citizens, recoup investments, make profits, and perform better than competitors
	Negative	Negative economic impact will be too large, Arabs will not contribute positively to economy	Needs to defend choices and avoid risks, uncertainty, and unknowns
Acceptability	Positive	Jewish paradigm will survive and "chosen people" will prosper, ancient mandate will be reclaimed, dispersion will be over	Needs to fit into tradition, communicate story, fulfill obligations, and establish social order
	Negative	Will remain outcasts, consensus will not be reached, and opponents will not be fair	Needs to avoid fears of rejection and fears of other paradigms
Validity	Positive	Land and sovereignty will be secure, civil and political rights will be protected	Needs for society to survive, security, and stability, needs to avoid risks, uncertainty, and unknowns
	Negative	World is hostile, conflict is inevitable, and Arabs will persist in trying to destroy Israel	Needs for protection and justification of actions

FIGURE 9.4. NBP Template for Israelis.

ATTRIBUTE	PERCEPTIONS	BELIEFS	NEEDS
Viability	Positive	Will contribute to meeting basic needs and open Israeli market	Needs for food, clothing, shelter, and education
	Negative	Positive economic impact will be too small, Israelis will not share prosperity	Needs to defend choices and avoid risks, uncertainty, and unknowns
Acceptability	Positive	Ethnic identity will be enhanced, mandate will be established, leadership will be recognized as legitimate, dispersion will be over	Needs to be socially redeemed, belong, fulfuill obligations, and communicate stories
	Negative	Will remain outcasts, consensus will not be reached, and opponent will not be fair	Needs to avoid fears of rejection and fears of other paradigms
Validity	Positive	Land and sovereignty will be gained, civil and political rights will be gained	Needs for safety, security, and stability, needs to avoid risks, uncertainty, and unknowns
	Negative	World is hostile, conflict is inevitable, and Zionism will survive and expand	Needs for protection and justification of actions

FIGURE 9.5. NBP Template for Palestinians.

Hypothesize Beliefs

Entries in the beliefs columns of Figures 9.4 and 9.5 were determined by reviewing the discussion of the problem thus far, and by consulting Figures 3.6, 3.11, 4.3, and 6.5. Comparing the beliefs columns of these two templates, one sees a few, but not many, differences. Israelis, in this analysis, are contemplating giving up territory at some economic cost, but gaining political recognition and acceptance. The Palestinians are considering gaining territory at the cost of accepting the existence of Israel. Both are pursuing ethnic identity, economic benefits, and sovereignty over the land they control as a result of the agreement.

Hypothesize Needs

The needs columns of Figures 9.2 and 9.3 were derived from reviewing the discussion of the problem and consulting Figures 3.5, 3.7, 3.10, 4.2, and 6.4. The needs columns of these two templates are quite similar. An important distinction, of course, is Israel's need to survive as a nation vs. the Palestinians' need to establish national identity. Israel also has a need to fit into a very specific ethnic tradition, while the Palestinians—not the Arabs in general—need social redemption in terms of a recognized identity and existence.

Evaluate Hypotheses

Other than personal experiences and the references cited, I have no data with which to evaluate the veracity of the needs and beliefs hypothesized. In an actual negotiation process, these hypotheses could be tested using questionnaires, interviews, and other assessment methods. Role playing could be used to illustrate the impact of needs and beliefs on the perceptions of the conflicting parties, as well as solicit reactions to the needs, beliefs, and perceptions depicted.

Modify Alternatives

First of all, I think that the principles suggested by Jimmy Carter (1985) and Herbert Kelman (1987) should be accepted as a starting point. Thus, the existence of Israel is a given, as is sharing of the land, the right to national existence, and the legitimacy of both sides' leadership. Further, the use of negotiation techniques such as advocated by Raiffa (1982), for instance, single negotiating text, and those employed in shaping the Camp David agreement can be accepted as given.

The central problem, therefore, is transforming win–lose to win–win. One way to look at this conflict is quite simple: Israel wants it all and the Palestinians want it all. One of them can have it all, but only temporarily. Moreover, having it all imposes substantial costs that are nonproductive in the sense that resources are consumed but not created.

Based on the earlier discussion of alternatives, I suggest that the key is transformation of the problem to one that portends very substantial win–win possibilities. This requires broadening the scope of the negotiation. The most obvious way to broaden the negotiation is to include the economic well-being and prospects of the region as a whole as elements of the discussions.

Perhaps the biggest stumbling block is to get the parties involved to accept the above givens, participate in negotiations, and have confidence in the stability of the results. Reviewing the entries of the two NBP Templates, it seems to me that central to overcoming this barrier is downplaying both Zionism and anti-Zionism. The emphasis has to shift from exclusion to inclusion, from nations pitted against each other to building a Middle East community.

Modify Situations

This proposal may sound rather fanciful. Its realization certainly may take a long time, but hopefully less than it took for the conflict to ripen. By slowly, but surely, modifying the situation, I think that fancy can turn into fact.

The key is building trust and confidence, lack of which is frequently associated with negative perceptions in the NBP Templates. The question, therefore, is what actions each of the conflicting parties could take to gain the trust and confidence of the other parties. I think that the best way to answer this question is to have the parties ask each other just that.

The impetus for such discussions might be the recognition that all of the parties are central stakeholders in each other's destiny. History has shown that successive win–lose confrontations in the Middle East have consistently yielded lose–lose results, at least relative to what might have been achieved. By consciously and systematically building mutual trust and confidence, substantial win–win results are possible.

Plan Life Cycle

The agreement for resolution of this conflict needs to have a long time horizon. Many issues can only be resolved in principle in the short term, and in practice in the longer term. This is due, in part, to the necessity of building trust and confidence as the process evolves. A long-term plan is needed in order to shape all parties' expectations about what benefits and costs will accrue in what time periods. This plan should also include periodic reviews and options for renegotiation.

Summary

This section has focused on the conflict in the Middle East to illustrate the use of the NBP Model, Template, and Analysis Methodology. As noted in the introduction of this chapter, this application of the methodology is admittedly very speculative. Hopefully, the results are interesting. However, I hasten to emphasize that people more knowledgeable about the Middle East are likely to find my prescriptions simplistic. Nevertheless, this analysis does show how the methodology can be employed to address exceedingly difficult problems.

IMPLICATIONS OF ANALYSES

It is easy to see the roles of expectations, attributions, and mental models in the conflicts discussed in this chapter. People's expectations for progress have seldom been satisfied, or their expectations for the worse have been frequently fulfilled. Each side attributes the other side's behaviors to the worst possible motives, which often may be justifiable, but certainly is not always the case. People's mental models of economics, politics, and human relations often seem biased at best, and perhaps flawed relative to the true range of possible outcomes.

Also apparent is how people in these conflicts tend to interpret their myths and models as facts. As noted in Chapter 3, this is a natural tendency, whether the domain is culture and religion, or science and technology. It is probably unreasonable to expect the majority of people to overcome this tendency. However, it is crucial that political and spiritual leaders support perspectives that are more inclusive and recognize the inherent value of the multiplicity of myths and models in our societies.

In doing the background research on these two conflicts and visiting the regions many times, it was inevitable that I sometimes found myself viewing our own society and culture through the "glasses" of these other societies and cultures. It seems to me that American beliefs, values, business practices, and so on reflect psychological and social needs common to humankind, as well as the particular set of experiences we have had as the country has evolved. The same can be said of South Africa—as depicted in *The Covenant* (Michener, 1980)—and the Middle East as depicted in *The Source* (Michener, 1965). To an extent, I guess this reflects a social and cultural nature vs. nurture dichotomy, which harks back to the discussion of Wilson and Dawkin's arguments in Chapter 3.

From a more practical point of view, this observation serves to emphasize the extreme importance of understanding the context of political conflicts, including the history whereby the context emerged. At any particular point in time, the state of a conflict includes not only the current situation—the evolution of needs, beliefs, and perceptions up to that point also play a central role. An important catalyst for change is an understanding of this evolution.

SUMMARY

This chapter has illustrated the use of the NBP Model, Template, and Analysis Methodology for identifying catalysts for resolving political conflicts. By understanding the needs and beliefs that underlie conflicting perceptions that are often the basis for conflict, this knowledge can be used as a catalyst for change, as a means for enabling innovative resolution of conflicts.

Two case studies were used to illustrate the use of the methodology for addressing this type of innovation problem. While the two examples are real and based on personal experiences and research, it is important to emphasize that I do not claim to have more than scratched the surface of these complex conflicts. Their primary purpose was to illustrate the methodology.

REFERENCES

Avishavi, B. (1991). Israel's future: Brainpower, high tech–and peace. *Harvard Business Review*, November–December, 4–12.

Campbell, J. (1988). *The power of myth*. New York: Doubleday.

Carter, J. (1985). *The blood of Abraham: Insights into the Middle East*. Boston: Houghton Mifflin.

Fischer, A., and Albeldas, M. (1987). *A question of survival: Conversations with key South Africans*. Johannesburg: Jonathan Ball.

Friedman, T. (1989). *From Beirut to Jerusalem*. New York: Farrar, Straus, and Giroux.

Henry, J. S. (1990). Even if sanctions are lifted, few will rush into South Africa. *New York Times*, October 28, p. 5.

Janeway, E. (1990). The smart money says South Africa. *New York Times*.

Keller, B. (1992a). The bond begins to fray in South Africa's politics. *New York Times*, October 25, p. 5.

Keller, B. (1992b). Apartheid party gains new voters. *New York Times*, November 1, p. 8.

Kelman, H. C. (1987). The political psychology of the Israeli-Palestinian conflict: How can we overcome the barriers to a negotiated solution? *Political Psychology, 8*, 347–362.

Laue, J. H. (1992). Ten pointers for conflict resolution. *United States Institute for Peace Journal, 5*, 4.

Lewis, F. (1990). A role for South Africa. *New York Times*, February 2.

Meredith, M. (1988). *In the name of apartheid: South Africa in the postwar era*. New York: Harper and Row.

Michener, J. A. (1965). *The source.* New York: Random House.

Michener, J. A. (1980). *The covenant.* New York: Random House.

Neibuhr, H. R. (1929). *The social sources of denominationalism.* New York: Henry Holt.

Raiffa, H. (1982). *The art and science of negotiation.* Cambridge, MA: Harvard University Press.

Rouse, W. B. (1989). *Black and white. Address at meeting of South African business executives.* Bush Lodge outside of Pretoria, August 8.

Rouse, W. B. (1991). *Thoughts on South Africa: Thirty months gone and thirty months to go.* Working paper distributed to South African business leaders. June.

Susskind, L., and Cruikshank, J. (1987). *Breaking the impasse: Consensual approaches to resolving public disputes.* New York: Harper and Row.

Chapter **10**

Enabling Change

This chapter integrates the discussions of earlier chapters. Of more importance, this chapter considers the implications of the notions set forth in this book. Topics of particular interest include the value of the Needs–Beliefs–Perceptions (NBP) methodology, development of enterprise support systems, and broader aspects of human-centered design.

Figure 10.1 lists the themes of this book as introduced in Chapter 1. The themes have been elaborated in terms of the 10 case studies listed in Figure 10.2. These 10 case studies are representative of the four archetypical innovation problems described in Chapter 1, namely, understanding the marketplace, enabling the enterprise, settling sociotechnical disputes, and resolving political conflicts.

In each case study, a central task was exploration of the context-specific nature of viability, acceptability, and validity. Perceptions of measures in these three categories were shown to provide insights into the likelihood of selling products, changing organizations, or solving problems. Insights were also gained into possible changes of products, organizations, and solutions.

Often, understanding perceptions "on the surface" of viability, acceptability, and validity is sufficient for success. In many cases, however, insights are needed "below the surface." Such deep understanding can be essential for the types of innovation problems discussed in earlier chapters.

The central premise of this book is that relationships below the surface can be fruitfully characterized in terms of relationships among needs, beliefs, and perceptions. The case studies show how such relationships can be framed in a wide variety of contexts. Of special importance in this framing process is deep knowl-

- Understanding the nature of viability, acceptability, and validity in any particular context is key to selling products, changing organizations, or solving problems in that context.
- Understanding the relationships among needs, beliefs, and perceptions will provide deep understanding of viability, acceptability, and validity.
- Innovation can be enabled by designing products, systems, services, organizations, and solutions in general that, at the very least, satisfy needs and do not conflict with beliefs, and potentially facilitate the constructive evolution of needs, beliefs, and perceptions.

FIGURE 10.1. Central themes.

edge of the sources of needs and beliefs. In fact, it can reasonably be argued that a primary benefit of the methodology is the background information that it causes you to collect.

Insights below the surface are useful in at least two ways. First, by understanding needs and beliefs, you are in a much better position to create products, organizations, and solutions that satisfy needs and do not conflict with beliefs. Quite simply, if you understand the real underlying problem, you are more likely to solve it!

A second use of these insights concerns facilitating constructive evolution of

Number	Chapter	Topic
1	3	New Venture Formation
2	4	Safety of Nuclear Power
3	6	Marketing and Sales of Software Products
4	6	Transitioning Defense Technologies to Civilian Applications
5	7	Changing from Service to Product Markets
6	7	Changing from Defense to Nondefense Markets
7	8	Environmental Protection vs. Economic Development
8	8	Trade Disputes
9	9	South African Conflict
10	9	Middle East Conflict

FIGURE 10.2. Case studies.

needs, beliefs, and perceptions. Many of the case studies uncovered needs and beliefs that would be better off resolved and put behind us—for example, predominant needs for food and housing, or dominating beliefs caused by, based on, and engendering hostility. By understanding the nature and role of such needs and beliefs, it is much more likely that you will generate creative ideas for evolving beyond these constraints.

Not surprisingly, I think that the 10 case studies provide compelling support for the value of the NBP Model, Template, and Analysis Methodology. From a theoretical point of view, these case studies do not prove that the needs–beliefs–perceptions formulation is "correct." However, from a practical point of view, they show that this formulation is useful across a broad range of domains and problems.

The remainder of this chapter focuses on a variety of subtleties and their implications. The goal is to place the NBP formulation in a broader context. In this way, it is possible to consider implications that go beyond the types of analyses discussed in this book.

BELOW THE SURFACE

What are catalysts for change? In a general sense, catalysts are concepts and principles that can enable the innovations needed in our complex world. Note that word "catalyst" is linked with the word "enable." Catalysts enable innovations. They are not, in themselves, the needed innovations.

More specifically, the understanding gained from a needs–beliefs–perceptions analysis provides the catalysts for change. By understanding relationships among needs, beliefs, and perceptions, we are enabled to see how alternatives and situations can be modified so as to improve perceptions. The primary goal is to change perceptions.

In some cases, an additional goal may be to change needs and beliefs. For many problems, this additional goal would be inappropriate. However, for problems where basic needs are the source of disagreement, or dysfunctional beliefs are impeding agreement, the understanding resulting from an NBP analysis may enable seeing how needs and beliefs can be constructively evolved.

Beyond the needs and beliefs below the surface of perceptions, a variety of other insights are often gained. One type of insight concerns the distinction between symptoms and causes. A dispute or a conflict, for example, may only be the symptom of a more fundamental cause. What is apparently a religious conflict, for instance, may actually be a socioeconomic conflict that has been conveniently "packaged" as a religious dispute.

Another type of insight concerns hurdles vs. barriers. Hurdles are obstacles that can be cleared with sufficient energy and determination. Barriers are insurmountable obstacles. It is important to recognize when expenditures of energy and determination—which usually are scarce resources—will be wasted because an obstacle is a barrier. For example, a negative perception may be a barrier if, due to underlying needs and beliefs, it is impossible to change this perception. In such cases, a tactic might be to assure that other perceptions are sufficiently positive to outweigh the immovable negative perception.

A third type of insight involves problems vs. opportunities. Sometimes managing an enterprise, for example, can seem to be just one problem after another. You just get one thing "fixed" when something else goes wrong. Below the surface, however, some of these problems may be opportunities.

This distinction can be clarified by thinking in terms of concepts of organizational learning discussed in Chapter 7. Fixing problems is often the hallmark of single-loop learning. In contrast, recognizing that a problem, perhaps a crisis, is an opportunity for change is, hopefully, the beginning of an episode of double-loop learning. By analyzing problems in terms of features below the surface, you can greatly increase the possibility of recognizing such opportunities.

VALUE OF THE METHODOLOGY

For many people who are involved with the types of innovation problems discussed in this book, the NBP Model, Template, and Analysis Methodology may seem overly systematic and severely constraining. Further, the idea that a simple, 10-step methodology can contribute to resolving such difficult problems may seem hopelessly naive. In fact, to an extent, I agree with these assertions!

Central issues in such discussions should be the role and value of this type of methodology. I am convinced that this methodology should provide a nominal path for pursuing the analysis necessary for contributing to the resolution of innovation problems such as emphasized in this book. Creative insights can be prompted and organized by using this methodology, but such insights are not generated by the methodology. This is yet another reason for using the word "catalyst" in conjunction with this material.

I emphasize the word "nominal" to denote a default line of reasoning that you employ in the absence of compelling reasons for proceeding otherwise. Use of this methodology should never be a substitute for thinking. Thus, for instance, quite frequently I do not pursue the 10 steps of the methodology in order. Moreover, I often iterate and jump around among steps in the process of organizing what I know, or need to know, about the problem at hand.

Use of the methodology is also, in part, an art. Choosing the attributes of viability, acceptability, and validity is often far from straightforward. The process of inferring beliefs and needs also cannot be completely proceduralized. Often, there is much "cut and try," in an attempt to make sense of the behaviors exhibited by stakeholders, as well as any other information that is available. Furthermore, as illustrated in the case studies, hypotheses are frequently evaluated very informally.

To a great extent, "the proof is in the pudding." The purpose of employing the methodology is *not* to develop new theories of needs, beliefs, and perceptions. The goal, quite simply, is to solve the problems of interest. If, as a result of employing the methodology, progress is made in solving problems, then the methodology has value.

The 10 case studies discussed in this book illustrate the wide applicability of the model, template, and methodology. I have also pursued very informal uses of this material in an attempt to understand perceptions associated with public policy issues, community plans, and social activities. I think these applications led to useful insights. However, I hasten to note that much more experience is needed, by a broad range of users, before it can be concluded that the methodology is usable and useful for a wide variety of problems. I look forward to hearing of other applications and the insights gained.

ENTERPRISE SUPPORT SYSTEMS

The experiences reflected in this book, as well as those underlying *Design for Success* (Rouse, 1991) and *Strategies for Innovation* (Rouse, 1992), have caused me to realize that most enterprises are in great need of more support as they formulate and pursue their missions and visions. They need help in planning improved and new products, systems, and services. They need help in envisioning and planning the future nature of their enterprise. As this book illustrates, they need help in dealing with numerous types of change.

Integrating the concepts, principles, methods, and tools from all three books can provide a strong basis for what I call an Enterprise Support System (ESS). An ESS provides the types of help just listed. This support is provided in three ways.

One way is *training*. This training includes the material from the three books, as well as training in context-specific methods and tools. Training can be provided using books, seminars, simulations, and coaching on the job.

A second way is *aiding*. Aiding requires that an enterprise's processes be reconsidered and, to the degree necessary, redesigned to assure ease of learning, ease of use, and acceptance by the stakeholders involved. Aiding can be provided by modifying processes, improving supporting documentation, using computer assistance, and mentoring by more experienced people.

The third way that an ESS provides support is by *integration*. Training and aiding should be considered as complementary elements of an overall support system. These two elements need to be integrated in terms of technology, functionality, and content. Training and aiding are often approached in piecemeal fashion, frequently independent of each other. The result can be mutually counterproductive efforts. Integration can be the means whereby the full potential of training and aiding is realized.

How do you conceptualize and create an ESS? My experience is that you can use the methodology in *Design for Success* to determine what form of ESS is needed and how it should function. The methodology in *Strategies for Innovation* can be employed to reconsider the nature of your enterprise in light of the impending creation of an ESS. Finally, the methodology in this text can be used to both understand the changes that an ESS portends below the surface and how these changes can best be accomplished and accommodated.

I realize that this prescription may seem overly recursive. I am suggesting that you employ a set of methodologies to create an ESS that, in part, supports use of these methodologies. Using the same concepts and terms on several levels can make it difficult to maintain a clear sense of important distinctions. However, despite these difficulties, the possibility of employing these methodologies in this way illustrates the broad applicability of the overall approach.

HUMAN-CENTERED DESIGN

To conclude this chapter, as well as this series of three books, it is important to focus on the essence of the overall approach that I am advocating. Quite simply, I assert that human-centered design is the essence of the many concepts, principles, methods, and tools described and illustrated in these volumes. In this final section, I elaborate the notion of human-centered design.

First, we should revisit the definition. Human-centered design is a process of assuring that the concerns, values, and perceptions of all the stakeholders in a design effort are considered and balanced. The word "design" in this definition denotes the activity of determining the function and form of a product, system, service, program, or problem solution in general. Thus, I view design to be a central activity of all humans, not just those whose job title is "designer."

A central element of this definition is the term "stakeholders." Choice of this term reflects the need to consider all people who have a stake in the process and results of a design effort. Stakeholders usually include one or more of parties such as customers, users, maintainers, producers, investors, regulators, voters, and so on. If a design effort results in a "win" for all of the relevant parties, overall success is quite likely and perhaps guaranteed.

For all stakeholders, human-centered design focuses on humans' roles, aspirations, values, and perceptions, as well as the needs and beliefs that underlie these constructs. The depth of the analysis of these attributes depends on the extent to which the problem at hand is important and controversial. Nevertheless, solutions of even quite straightforward problems can benefit from at least informal reflection on principles of human-centered design.

For example, I find that the distinctions among viability, acceptability, and validity are useful to keep in mind during discussions in virtually any problem-solving group. When people express disagreement about the likely impact or desirability of a course of action, I try to assess the extent to which their concerns are predominantly in one of these three categories. I usually ask questions to evaluate my hypotheses. Quite commonly, apparent disagreements can be diagnosed and perhaps eliminated in this way.

Human-centered design can also be viewed much more broadly. Moving beyond the concept of design, the human-centered construct can be applied to human relations in general. My guess is that true "community," such as advocated by Scott Peck (1987) and others, can be fostered by maintaining a human-centered perspective. By always concerning ourselves with stakeholders and their concerns, values, and perceptions, we are likely to become catalysts for change and enable many innovations. The result will be an Enterprise Support System for the broadest enterprise of all.

REFERENCES

Peck, M. S. (1987). *The different drum: Community-making and peace*. New York: Simon and Schuster.

Rouse, W. B. (1991). *Design for success: A human-centered approach to designing successful products and systems*. New York: Wiley.

Rouse, W. B. (1992). *Strategies for innovation: Creating successful products, systems and organizations*. New York: Wiley.

Author Index

Subject Index